Advances in Geographic Information Science

Series editors

Shivanand Balram, Burnaby, Canada
Suzana Dragicevic, Burnaby, Canada

More information about this series at http://www.springer.com/series/7712

Chenghu Zhou · Fenzhen Su
Francis Harvey · Jun Xu
Editors

Spatial Data Handling in Big Data Era

Select Papers from the 17th IGU Spatial Data Handling Symposium 2016

Springer

Editors
Chenghu Zhou
State Key Laboratory of Resources and Environmental Information System, Institute of Geographical Sciences and Natural Resources Research
Chinese Academy of Sciences
Beijing
China

Fenzhen Su
State Key Laboratory of Resources and Environmental Information System, Institute of Geographical Sciences and Natural Resources Research
Chinese Academy of Sciences
Beijing
China

Francis Harvey
Leibniz Institute for Regional Geography
Leipzig, Sachsen
Germany

Jun Xu
State Key Laboratory of Resources and Environmental Information System, Institute of Geographical Sciences and Natural Resources Research
Chinese Academy of Sciences
Beijing
China

ISSN 1867-2434 ISSN 1867-2442 (electronic)
Advances in Geographic Information Science
ISBN 978-981-10-4423-6 ISBN 978-981-10-4424-3 (eBook)
DOI 10.1007/978-981-10-4424-3

Library of Congress Control Number: 2017937118

Printed on acid-free paper

This Springer imprint is published by Springer Nature
The registered company is Springer Nature Singapore Pte Ltd.
The registered company address is: 152 Beach Road, #21-01/04 Gateway East, Singapore 189721, Singapore

Preface

The 17th International Symposium on Spatial Data Handling (SDH) was held on 18–20 August 2016 in Beijing. As one of the oldest GIS conferences, this symposium aims to bring together geographers, cartographers, computer scientists and others from the international community of geospatial information sciences and geomatics engineering to present the latest achievements and to share experiences in GIS research.

The International Symposium on Spatial Data Handling (SDH) is the biennial international research forum for Geospatial Information Science (GIScience), co-organized by the Commission on Geographic Information Science and the Commission on Modelling Geographical Systems of the International Geographical Union (IGU). It commenced in 1984 in Zurich, Switzerland and has ever since been held in Seattle, USA; Sydney, Australia; Zurich, Switzerland; Charleston, USA; Edinburgh, UK; Delft, The Netherlands; Vancouver, Canada; Beijing, China; Ottawa, Canada; Leicester, UK; Vienna, Austria; Montpellier, France; Hong Kong, China; Bonn, Germany and Toronto, Canada.

This is the second time SDH hold in Beijing, China (last time SDH 2000), again hosted by the State Key Laboratory of Resources and Environmental Information System (LREIS), the Institute of Geographic Sciences and Natural Resources Research (IGSNRR), Chinese Academy of Sciences. This year, the theme of the symposium is "face the challenges of big data within GIS". During two and half days of conference, the participants introduced their recent work and discussions were held on the storage, retrieval, visualization and knowledge discovery of spatial big data; multi-scale spatial data representations; data-intensive geospatial computing, and space-time modeling in open source environment; quality issues, spatial analysis and geographical applications of spatial data.

For SDH 2016, over 65 paper proposals were received and reviewed by the international program committee and additional reviewers. Based on the peer review of abstracts, 54 papers have been accepted to be presented in the conference, in which 35 are recommended to be developed to full papers. Finally, 20 papers were submitted to the second round of peer review, and 15 papers were selected to be included in the proceedings. For convenience, they were classified into four topics, though some papers may be involved in more than one topic.

The growing volume of geo-referenced data has caused increasing needs of efficient storage, processing and retrieval of the massive data. In the first part of the SDH 2016 proceedings, Data Intensive Geospatial Computing and Data Quality, Mudabber Ashfaq and co-authors test the performance of spatiotemporal queries on T-drive and BerlinMOD trajectory datasets in Parallel Secondo environment. The results of their experiments indicate that the optimal number of nodes depends on the volume of data and the complexity of the query. Peng Wang and co-authors introduced their work of integrating a few hundred GB of multidisciplinary data in a system for geological disposal of high-level radioactive waste, including geo-information model design, metadata management development, and data management system implementation. Uncertainty and error are still unavoidable issues in Big Data Era. In the paper of Bo Sun and co-authors, they deem that the ground reference used as ground truth in validation and assessment of remote sensing land-cover classification contains errors, especially those coming from big geographic data. By cross-validation of ground references sampled by image interpretation and filed investigation, they find the uncertainties in ground reference data, and suggest the need for new evaluation system. Taking the error in Predictive Vegetation Mapping (PVM) as an example, Brian Lees discusses problem of keeping balance between accuracy and cost of efficiency when visualizing spatial data in his paper, and points out that some types of errors are acceptable for the convenience of representation.

In the second part, Web and Crowd Sourcing Spatial Data Mining, Shuang Wang and co-authors introduce an Emergency Event Information Extraction System. With the help of knowledge base and learning patterns form examples, the system can automatically extract event information from unstructured online news and save it in machine-readable database. Hou and Murayama describe an approach to evaluate people's utilitarian walking behavior using People Flow Data, and compared the result with neighborhood environment acquired with multi-criteria evaluation of residential density, street connectivity, land-use density, bus stop density and railway station accessibility. Their work provides a method to study neighborhood environment at metropolitan-level. Aiming at analysis of large tracking datasets of moving objects, Hongbo Yu proposes concepts of space-time path and station to model trajectories and locate spatial and temporal cluster of paths. He implements these concepts and some aggregation methods in a 3D space-time GIS environment, and the spatiotemporal patterns of large trajectory dataset are explored and visualized at different levels in the space-time GIS environment. The flow data can form network connecting different locations. To detect the interaction of locations, Zhixiang Fang and co-authors propose an extended community detection algorithm based on CNM algorithm to find communities constrained by already partitioned source nodes. With human mobility flows derived from mobile phone data in Shenzhen City, they find the human communities and compare them with the planned urban polycentric clusters.

The third part, Visualization of Big Geographical Data, points to the representation of complex patterns and processes of GIScience. Addressing the theoretical and representational limits of cartographic visualization in scientific research, Francis Harvey suggests a new methodology for GIScience visualization that rests on transformational approach. A heuristic approach and the process to build the representation are introduced and the conceptual framework and foundation of the methodology are explained. From the perspective of visual analytics, Alan MacEachren discusses the importance and feasibility of understanding and constructing place from unstructured big data. The concept of place in different dimensions and the "5Vs" character of big data are explicated, and some (geo)visual analytical tools concerning three of the five "Vs" to leverage big data are introduced. Xun Wu and co-authors deal with the generalization of road network. To decide the roads that should be selectively omitted, they analyze four structure indices and one geometric index of road network and find a composite index to define the importance of the roads. The composite index is then validated with a road removing approach. The paper from Xiaoqiang Cheng and co-authors deals with the visualization of VGI data on different devices. Based on the scale-dependent feature of coalescence, they propose an index called degree of coalescence to measure the degree of visual coalescence, and use it in a detail resolution model which describes the level of detail information of vector lines. The model can be used in map generalization and overcome the scale heterogeneity of VGI.

The three papers in the final part of the proceedings, Spatial Analysis and Simulation, all point to the urban issues. Nowadays, a lot of approaches have been explored in data rich environment. However, sometimes we still are faced with data limitation. Shyamantha Subasinghe and Yuji Murayama develop an Urban Growth Evaluation Approach (UGEA) to detect urban growth in data poor situation by integrating the neighborhood interactions of urban land-use categories. Masahiro Taima, Yasushi Asami and Kimihiro Hino examine the city blocks with only office buildings to predicate the building footprint and office shape in Tokyo. A building location estimation (BLE) model is developed to estimate building locations based on the shape of a city block, and the probability of building coverage for each point on every floor is visualized as a spatial image. Cellular automata (CA) is a widely used approach for spatial simulation of land-use and land-cover changes. Wenyou Fan and co-authors adopt the CA SLEUTH model to simulate the urban expansion of Wuhan City, and predict future urban boundary in different scenarios.

Finally, we would like to express our appreciation to all those who have submitted their research and attended the meeting. Your participation made SDH 2016 successful. We also thank the program committee and additional reviewers for reviewing and sharing their experience, and thank the steering committee for their support. Thanks also go to the local organizing committee, and the staff and student volunteers in LREIS.

Beijing, China	Chenghu Zhou
Beijing, China	Fenzhen Su
Leipzig, German	Francis Harvey
Beijing, China	Jun Xu
December 2016	

Organizing Committee

General Chair

Fenzhen Su, LREIS, Chinese Academy of Sciences
Francis Harvey, Leibniz Institute for Regional Geography

Program Chair

Chenghu Zhou, LREIS, Chinese Academy of Sciences
Brian Lees, University of New South Wales
Jiuling Sun, LREIS, Chinese Academy of Sciences
Anthony G.O. Yeh, The University of Hong Kong

Organizing Committee

Huayi Wu, Wuhan University
Tao Pei, LREIS, Chinese Academy of Sciences
Chengzhi Qin, LREIS, Chinese Academy of Sciences
Jun Xu, LREIS, Chinese Academy of Sciences
Jie Chen, LREIS, Chinese Academy of Sciences
Jianghao Wang, LREIS, Chinese Academy of Sciences

Program Committee

Benenson Itzhak, Israel
Francis Harvey, Germany

Juri Roosaare, Estonia
Nadine Schuurman, Canada
Åke Sivertun, Sweden
Andrej Medvedeff, Russia
Lars Brabyn, New Zealand
Pavlos Kanaroglou, Canada
Graham Clarke, UK
Manfred M. Fischer, Austria
Daniel A. Griffith, USA
Yukio Sadahiro, Japan
Jinfeng Wang, China
Huayi Wu, China
Qiming Zhou, Hong Kong
Cunjian Yang, China
Xiang Li, China
Qing Zhu, China
Xiaoping Liu, China
John Wilson, USA
Jean-Paul Bord, France
Stan Geertman, The Netherlands
Mauro Salvemini, Italy
Abdul Rashid Bin Mohamed Shariff, Malaysia
Vit Vozenilek, Czech Republic
Shoaib Khalid, Islamic Republic of Pakistan
Stewart Fotheringham, USA
Ali Bennasr, Tunisia
Danny Czamanski, Israel
Boris I. Gartsman, Russia
Beniamino Murgante, Italy
Therese Steenberghen, Belgium
Vladimir Tikunov, Russia
John Shi, Hong Kong
Yu Liu, China
Diansheng Guo, USA
Min Deng, China
Zhilin Li, China
Huiping Liu, China

Contents

About the Editors

Chenghu Zhou received his Ph.D. from the Institute of Geographical Sciences and Natural Resources Research, Chinese Academy of Sciences, with a focus on cartography and GIS. He is currently an Academician at the Chinese Academy of Sciences.

Fenzhen Su completed his Ph.D. in GIS and Cartography at the Institute of Geographical Sciences and Natural Resources Research, Chinese Academy of Sciences, Beijing. He is currently Director of the State Key Lab of Resources and Environmental Information Systems, Institute of Geographical Sciences and Natural Resources Research, Chinese Academy of Sciences, Beijing, China.

Francis Harvey completed his Ph.D. at the University of Washington, Seattle, Washington. He has been head of the Department of Cartography and Visual Communication, Leibniz Institute for Regional Geography, since 2015.

Jun Xu received his Ph.D. in Geographical Information Systems from the Department of Geography, State University of New York at Buffalo. Her research interests are in the fields of geographical ontology, spatial knowledge representation and qualitative reasoning, and spatial data mining. She is now Associate Professor at the State Key Laboratory of Resources and Environmental Information Systems, Institute of Geographical Sciences and Natural Resources Research, Chinese Academy of Sciences, Beijing, China.

Part I
Data Intensive Geospatial Computing and Data Quality

Using T-Drive and BerlinMod in Parallel SECONDO for Performance Evaluation of Geospatial Big Data Processing

Mudabber Ashfaq, Ali Tahir, Faisal Moeen Orakzai, Gavin McArdle and Michela Bertolotto

Introduction

With the growing availability of ubiquitous mobile computing devices such as smart phones equipped with GPS, the amount of mobility data is increasing. Typically mobility data contains both spatial and temporal data which form trajectories representing a time-stamped path of an object through space. Movements of these objects contain hidden patterns which reflect the behavior of these entities. Spatio-temporal queries are commonly used to identify such patterns. While there are several DBMS which provide support for spatial operators, only few specialized ones provide support for both spatial and temporal data processing. Secondo (Guting et al. 2005) and Hermes (Pelekis et al. 2006) are two examples. The nature

M. Ashfaq · A. Tahir (✉)
Institute of Geographical Information Systems, National University of Sciences and Technology, Islamabad, Pakistan
e-mail: ali.tahir@igis.nust.edu.pk

M. Ashfaq
e-mail: mudabber13@igis.nust.edu.pk

F.M. Orakzai
Department of Computer Science, Aalborg University, Aalborg, Denmark
e-mail: fmo@cs.aau.dk

G. McArdle · M. Bertolotto
School of Computer Science and Earth Institute, University College Dublin, Dublin 4, Ireland
e-mail: gavin.mcardle@ucd.ie

M. Bertolotto
e-mail: michela.bertolotto@ucd.ie

C. Zhou et al. (eds.), *Spatial Data Handling in Big Data Era*, Advances in Geographic Information Science,
DOI 10.1007/978-981-10-4424-3_1

of movement data means that its size can become very larger, and processing and querying them become slow and inefficient. To handle such instances, there are moving object database platforms which support parallel query processing.

Handling big data requires high performance computing or distributed data processing. The state-of-the-art industrial standard is the MapReduce model (Dean and Ghemawat 2008). The framework of Apache Hadoop (Murthy et al. 2011) is its open-source implementation. The original aim of the MapReduce paradigm was to process simple text documents. However the implementation of complex algorithms and the management of heterogeneous data structures was a challenging task. To counteract this, several extensions and toolkits have been introduced that operate over the Hadoop platform enabling a wide range of data management, mining and analysis possibilities.

Parallel Secondo, a Hadoop based platform is a promising tool to handle big mobility data. It combines the distributed processing ability of Hadoop and the useful analytical capabilities of Secondo to store and process trajectories. Parallel Secondo provides hybrid processing where analysis can be run on both sequential and parallel modes depending on the available distributed architecture. Data size also drives the necessity to run queries in sequential or parallel mode. Processing queries using more than one node may increase time efficiency; however the extent of efficiency increase includes many factors such as volume of data, nature of query and number of nodes. This paper proposes an appropriate environment to achieve efficiency by using the optimum amount of computation power which can reduce cost.

The main focus of the study is to evaluate the response of a parallel system when the volume of data differs along with a type of spatio-temporal queries. The study uses Parallel Secondo. A step-wise performance has been evaluated in trajectory analysis using both sequential and parallel modes. For the experiments, a sample *T-drive* dataset of taxi movement in Beijing has been used. This dataset consists of 10,357 taxi trajectories containing approximately 15 million points. Similarly, the dataset *BerlinMOD* contains approximately 300,000 synthetic car trips in Berlin. This dataset also has a set of predefined spatio-temporal queries specifically designed to test and benchmark spatio-temporal DBMS.

The remainder of the paper is organized as follows. Section “Related Work” gives an overview of related work in geospatial big data and mobility analysis. The methodology is presented in Section “Methodology” while Section “Results and Discussions” presents the results of our study along with a discussion of the main findings. Finally the conclusions and direction to future work are given in Section “Conclusions and Future Work”.

Related Work

Lee and Kang (2015) discussed the challenges and opportunities of big data. Of the 15 global challenges faced by humanity they consider, 7 challenges are directly related to spatial information acquisition and planning. These challenges include energy, clean water, health, population and resources, transnational organized crimes, and peace and conflict. These challenges require the analysis of massive spatial datasets and analytical information.

Recognizing the role of spatial data, the framework of Hadoop (Murthy et al. 2011) has been adapted and extended to process big geospatial data and several efforts in this area are ongoing. For example, ESRI, a supplier of Geographic Information Systems has added Hadoop support in their ArcGIS products via the ESRI Geometry APIs which spatially enable the Hadoop cluster for scalable processing of geo-tagged data. Using this approach Gao et al. (2014) used Volunteered Geographic Information (VGI) sites and automatically linked the results with ArcGIS Desktop for visualization. The authors further constructed a gazetteer from the geospatial information collected from social media such as Flickr through geo-tagged images. Spatial Hadoop is an example of another extension of Hadoop to process spatial data (Eldawy 2014).

Spatial On-Line Analytical Processing (SOLAP), which is a powerful decision support system for exploring the multidimensional perspective of spatial data, was introduced by Li et al. (2014) for environmental monitoring which requires spatio-temporal analysis. The author used this system with the Hadoop MapReduce framework for processing large remote sensing data. Similarly, in another study, the feasibility of Hadoop MapReduce in a parallel model was explored with a broader domain of scientific data and installed in a cluster of high-end computers (Golpayegani and Halem 2009). In their research the authors proposed that Hadoop may be used to improve scientific satellite data processing time.

All the Hadoop based big data platforms mentioned above can handle typical vector and raster datasets. However, there are very few big data platforms which can support moving objects such as trajectories. Parallel Secondo is one such system which is an open source moving object database developed by Lu and Guting (2014). According to the authors, Parallel Secondo is useful for handling big mobility data since it provides Secondo functionality in a distributed environment. Parallel Secondo uses Hadoop for parallel processing. Lu and Guting (2014) compared the processing efficiency of Secondo and Parallel Secondo. They performed different queries on OpenStreetMap data as well as GPS data of personal trajectories. In a similar study, Orakzai (2014) compared the time efficiency of Parallel Secondo to his own approach for processing big mobility data in a parallel environment using HBase and achieved better performance than Parallel Secondo. However, the system is only a prototype and is not available for general public to experiment.

There are several studies which used movement data sets such as *T-drive*. Yu et al. (2015) proposed a system called Cludoop and used *T-drive* with the Hadoop MapReduce based algorithm for clustering the large dataset. A comprehensive experimental evaluation on 10 network-connected commercial PC's was carried out using both huge-volume real and synthetic data. The authors demonstrated the effectiveness of the algorithm in finding correct clusters with arbitrary shape and the scalability of their proposed algorithm was better than state-of-the-art methods.

Furthermore they compared the effectiveness of the algorithm against other density-based clustering algorithms.

Other studies used *T-drive* to demonstrate knowledge discovery. For example, Liu et al. (2011) proposed an effective method to detect outliers from the taxi trajectory dataset of Beijing. The authors also looked into the interaction between taxis. Similarly, Zhu et al. (2012) inferred the status of the taxi. For instance, information was available such as whether a taxi was occupied, available or parked. They used probabilistic decision trees and trained the classifiers followed by Hidden Semi Markov Model (HSMM) for long term travel patterns. In another study by Yuan et al. (2011), the authors proposed a recommender system for people and taxi drivers using *T-drive* and incorporated routing. Other trajectories datasets, for example generated by Human Computer Interaction, were used for spatial recommendation and map personalization (Tahir et al. 2014; McArdle et al. 2015). In all of the above mentioned studies, authors discuss their own approach without recommending any big data platforms.

The use of open source geospatial big data platforms such as Parallel Secondo is a good choice for running and executing realtime analysis. Moving object datasets such as *T-drive* and *BerlinMOD* are benchmarks which are tested with Parallel Secondo in our study.

Methodology

This section describes the datasets, the geospatial big data platform, and the machine specifications on which the experiments were tested.

Datasets

To conduct our experiments, *T-drive* and *BerlinMOD* datasets were selected. *T-drive* is a real taxi movement data while *BerlinMOD* is a synthetic dataset of vehicle movements. The data in these collections are different. *T-drive* data consists mainly of spatio-temporal sequences while *BerlinMOD* has more attributes information such as vehicle type, vehicle license number and vehicle trips etc. The detailed descriptions of the datasets are as follows.

T-Drive

T-drive dataset was created by Microsoft Research Asia. The average sampling interval is about 177 s with a distance of about 623 m. The data used in this research are taxi tracks of 7 days only. The size of T-drive dataset is 1.6 GB. This does not qualify for big data however the possibility to scale up the analysis provides an insight into the integrity of the system. Full details and properties of *T-drive* are given below:

- Time Span of the Collection: 02/02/2008–08/02/2008
- Number of Users: 10,537
- Number of trajectories: 10,359
- Number of points: 15 million
- Total distance: 9 million km
- Total duration: 144 h.

BerlinMOD

BerlinMOD (Duntgen et al. 2009) benchmark designed at University of Hagen and is used to measure the queries performance on moving objects data such as trajectories. The data has been sampled from the moving point data of cars using driving simulations on the street network of Berlin. Furthermore the simulations reflect the behavior of workers who commute between their work locations and homes. There is a data generator available which can be used to generate data as needed and by specifying the required parameters. The data is generated with real world spatial data that can be imported using a tool called bbike.[1] This tool contains real spatial data consisting of Berlin streets.

BerlinMOD is just a collection of dataset and benchmark queries. It does not scale the database. It contains queries for both Secondo and Parallel Secondo. For experimental purposes, a scale factor of 1.0 was used to find the efficiency of Parallel Secondo in multiple nodes. A parallel generator is specifically prepared for creating a large amount of benchmark data for the evaluation. It contains several Secondo scripts and a Hadoop program. The total size of *BerlinMOD* is 1.7 GB while it contains 2000 trajectories and attribute information like vehicle registration, number and type etc.

Parallel Secondo

Parallel Secondo allows the user to handle non-standard data and applications especially those containing spatio-temporal information (Lu and Guting 2014). This database system was developed for processing mobility data. In addition, Secondo could be utilized for handling other data types such as vectors and rasters. Secondo is an open source system and its source code is freely available for users to access and modify according to their requirements and environment.

[1]http://bbbike.de.

Parallel Secondo queries are processed on MapReduce standard. The queries are first converted into Hadoop jobs by the master database. Hadoop further divides the job into map and reduce tasks. The tasks are processed in parallel on all slave data servers. Each task fetches its essential data from either the local slave database or remotely from the other computers through Parallel Secondo File System (PSFS). PSFS is a simple distributed file system which is used to minimize the migration overhead among Hadoop and slave databases.

Hadoop based Parallel Secondo was selected for our study as it provides users with the variety of spatio-temporal algebras that are useful for analysis of mobility data. With the addition of Hadoop, the tool provides the ability to perform in a parallel environment. The Hadoop framework assigns program tasks running on a computer cluster in parallel. A task's embedded procedures are managed by Secondo for the sake of efficiency. From Secondo, Parallel Secondo inherits the capability of processing spatial and moving objects.

Hardware Computer System

For performance evaluation two systems were used. Both systems had powerful computation abilities with multiple processors and were able to process big data. The specifications of a single node system are as follows:

- Processor Intel Core i7-4790 CPU @ 3.60 GHz 8
- Random Access Memory (RAM) 8 GB
- OS Ubuntu 14.04 LTS 64 bit.

The specifications of a multi-node system configured with Parallel Secondo are as follows:

- PowerEdge R720/R720xd Motherboard TPM
- 2× Intel Xeon E5-2680v2 2.8 GHz, 25 M Cache, 8.0GT/s QPI, Turbo, HT, 10C, 115 W, Max Mem 1866 MHz
- 256 GB (16 × 16 GB) RDIMM, 1600 MHz, Low Volt, Dual Rank, ×4 Bandwidth.

Results and Discussions

This section discusses the performance of Parallel Secondo when running queries on samples from *T-drive* and *BerlinMOD* datasets.

T-Drive Performance

To query the *T-drive*, the data were divided into four parts (each consisting of 1, 5, 10 and 15 million points respectively). Each part is run with multiple nodes to observe the optimal performance. Parallel Secondo was configured on a single node as well as on multiple nodes. Spatio-temporal queries were selected based on filtering, clustering, search and creating trajectories from points. These queries are basic steps often required when analyzing mobility data. All spatio-temporal queries were run with 1, 5, 10 and 15 million points.

Figure 1 illustrates the minimum amount of time required to process creating trajectories query in Parallel Secondo distributed environment. The x-axis shows number of data points while y-axis plots the time duration. The graph demonstrates the performance of a single node configured with Parallel Secondo. It is argued that parallel queries are efficient only for big data and for small data conventional sequential queries should be used.

This can be seen in Fig. 1 where efficiency and processing time only improve after the maximum number of points (15 Million) were used. This shows the usefulness of a system even on a single node. However this is not a significant difference for querying 15 million points in parallel and sequential mode.

On the other hand, increasing nodes in a distributed system for optimal performance is not always the best solution. This may decrease the efficiency of the processing (see Fig. 2). For this experiment, 10 nodes were configured with Parallel Secondo. Each part of data was evaluated against all 10 nodes. The graph demonstrates that increasing the number of nodes indefinitely does not produce the optimal results. This is due to the additional overhead created by the time required to distribute the job to different nodes. Therefore this will decrease the efficiency after achieving the optimal performance. Hence nodes should be assigned according to the size of the data in order to achieve the best results.

Additionally there are other factors which can cause slow distribution of jobs. For instance, one factor is the cost of join. When the datasets need to be joined and stored on different nodes, the data need to be transmitted between the nodes to perform the join. Network bandwidth can also make distributed computing slower. Orakzai et al. (2015) describe partial parallelization, partial cluster utilization, network cost, disk IO, and sorting costs among others which can reduce the performance.

Figure 3 depicts the efficiency of the parallel system against the conventional sequential methods. The x-axis shows the number of data points along with the respective nodes which provide the maximum efficiency for particular size of dataset. 1 million data points are not big enough to be processed in parallel environment hence sequential query would be a better approach for this data size. However 2 nodes will give the maximum efficiency in parallel environment for 1 million points. Similarly for 5 million data points, the ideal performance of Parallel Secondo is 3 nodes while for 10 million points, it is 4 nodes. For 15 million points, our study shows that 6 nodes produce efficient results. Due to its hybrid nature,

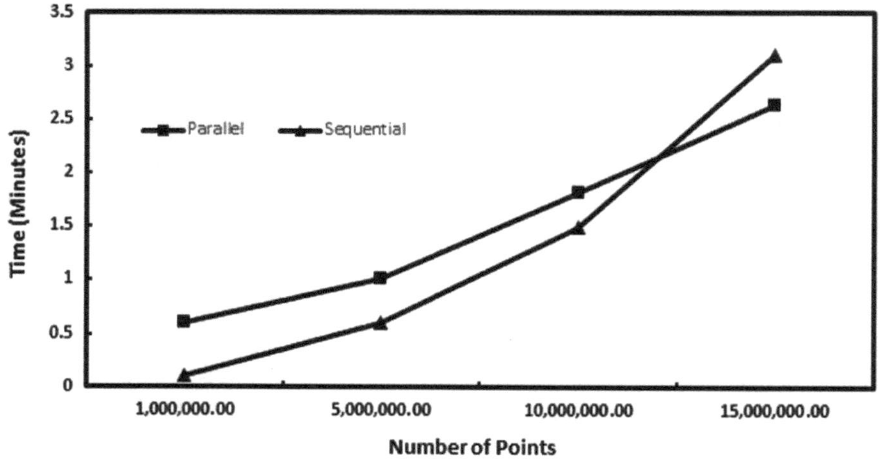

Fig. 1 Computational efficiency of a single node on *T-drive*

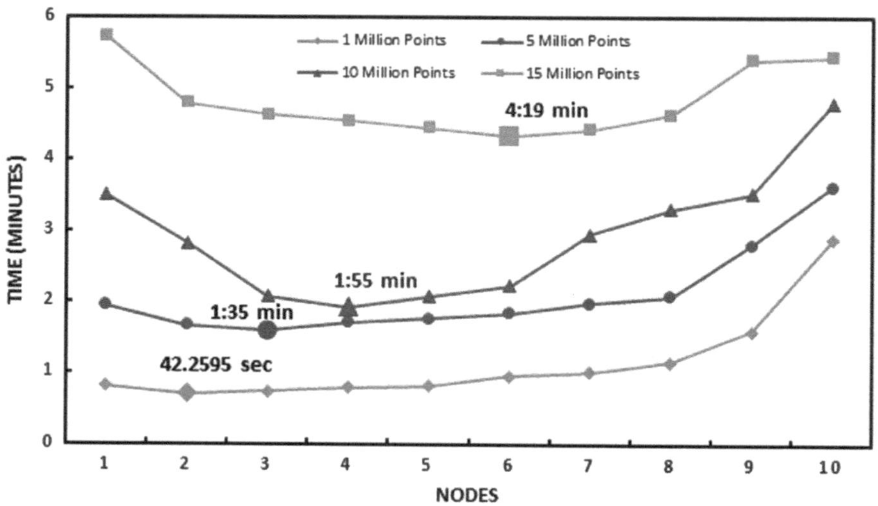

Fig. 2 Computational efficiency of multiple nodes on T-drive

Parallel Secondo can process both sequential and parallel queries. This makes it a powerful geospatial big data platform. The main point here to stress is that the right parallelization factor can lead to a linear scaleup. When the cost of parallelization exceeds the parallelization benefit, the performance starts degrading.

Efficient performance not only depends on the data size and number of nodes available in the parallel environment, but also on the type of spatio-temporal operator being used in the queries. In our case, we ran the queries using filter, clustering and search operations. Figure 4 illustrates the result of a filter query.

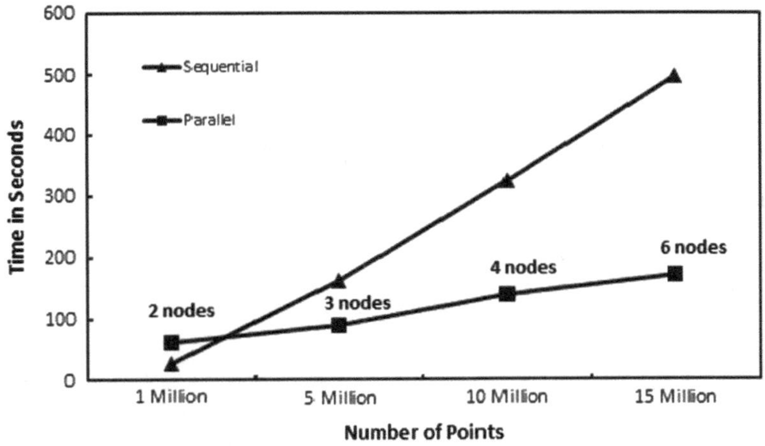

Fig. 3 Trajectory creation query on both sequential and parallel modes

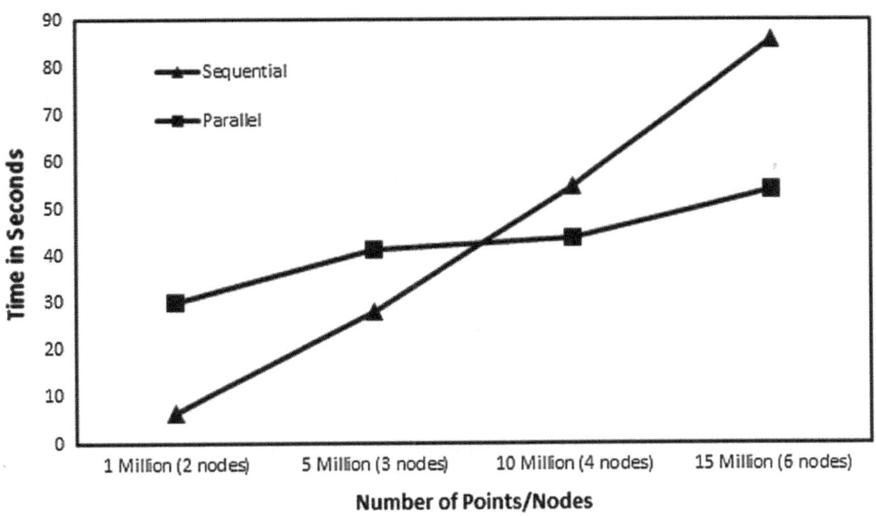

Fig. 4 Filter query on both sequential and parallel modes

A query was performed to retrieve the value of some of the attributes of a particular taxi.

Similarly, Fig. 5 shows the results of all queries when run in both sequential and parallel mode. The queries which are used in experiments include calculating the speed, time, direction and length of each trajectory. For optimal performance, some of these queries can be merged into a single query.

The clustering query was also applied to the *T-drive* dataset. The Density Based Spatial Clustering of Applications with Noise (DBSCAN) algorithm was used to find

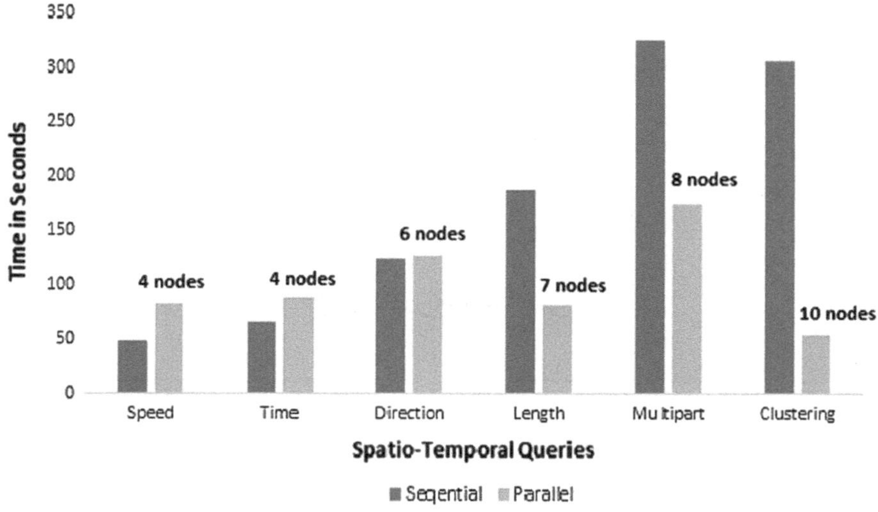

Fig. 5 Computational efficiency of multiple nodes on both sequential and parallel on *T-drive*. The optimal number of nodes are also shown

the similar trajectories based on a given distance threshold. Clustering is a computationally expensive operation. It performed better in parallel environment by reducing the significant time difference (see Fig. 5). Similarly, length and combined operations outweighed sequential processing in Parallel Secondo environment. These results help analysts to identify the type of operations when working on big data platforms.

BerlinMOD Performance

In the previous section, the *T-drive* dataset was used to run experiments. This data has limited attribute information and so limits the complexity of the spatio-temporal queries which can be tested. Therefore generalized spatio-temporal queries were run on trajectories. *BerlinMOD* includes rich attribute information which enables some specialized query analysis. The detailed queries we ran are discussed in (Duntgen et al. 2009). *BerlinMOD* uses object based and trip based representations of the created movements. These representations provides range style and nearest-neighbors queries. For our experiments we used range-style queries which deal with spatial, temporal and spatio-temporal predicates. These queries were configured and run in the Parallel Secondo environment.

Performing these queries enabled us to analyze the performance of Parallel Secondo on the *BerlinMOD* dataset. The same approach is used in *BerlinMOD* to test which queries are better with sequential or parallel modes. Figure 6 illustrates the performance of each query both in parallel and sequential modes. The x-axis shows all the queries of *BerlinMOD* while the y-axis shows the time duration in seconds.

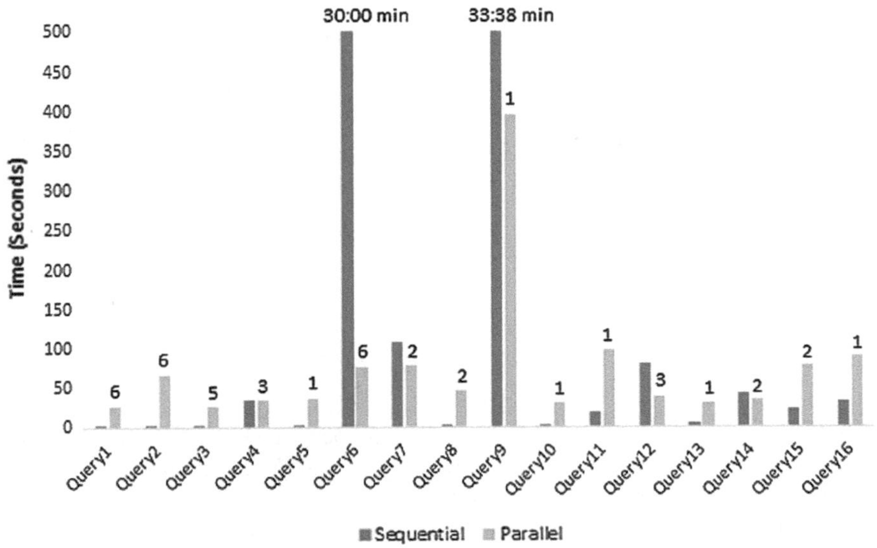

Fig. 6 Computational efficiency of multiple nodes on both sequential and parallel queries of *BerlinMOD*. The optimal number of nodes are also shown

The graph displays the effectiveness of parallel computing on different spatio-temporal queries. According to this graph, queries 6, 7, 9, 12 and 14 show better performance of the parallel environment whereas all other queries performed better with sequential mode. Some of the queries did not perform well in parallel environment due to the fact that the dataset was not sufficiently big (1.7 GB in volume). Query 8 contains distance operation on filtered trajectories. This operation makes it inefficient for parallel processing.

In the analysis below, we evaluate the performance of *BerlinMOD* queries in Parallel Secondo. Query 1, 2 and 3 did not show good performance, while Query 6 performed the best in Parallel Secondo.

Query 1

What are the models of the vehicles with license plate numbers from Querylicenses?

The query is written in sequential as:

```
let SQ_OBACRres001 = Querylicenses feed {O}
loopjoin[ dataSCcar_license_btree
dataSCcar exactmatch[.license_O]]
project[license, Model]
consume;
```

The query is written in parallel as:

```
let OBACRres001 =
Querylicenses_Dup_List
hadoopMap[ "Q1_Result", DLF; . {O} loopsel[
para(dataSCcar_license_btree_List)
para(dataSCcar_List)
exactmatch[.license_O]]
project[license,Model]
]
collect[] consume;
```

In the parallel query, *hadoopMap* is the main operator which assigns jobs to all available nodes. The map job is performed by MapReduce algorithm and a loop extracts models of specified license plates with the help of *exactmatch* operator. On the other hand, sequential mode does not go through all these transformations and hence simple loop and *exactmatch* operators show high performance with low processing time. We note that the use of additional operators increases the time of processing and thus decreases the efficiency.

Query 2

How many vehicles are passenger cars?

The query is written in sequential as:

```
let SQ_OBACRres002 = dataSCcar feed filter [.Type = "passenger"]
count;
```

The query is written in parallel as:

```
let OBACRres002 = dataSCcar_List
hadoopMap[ "Q2_Result", DLF; .
feed filter[.Type = "passenger"]
]
collect[] count;
```

Query 2 lacks extra operators which may degrade the performance of a parallel query. As discussed for the *T-drive* dataset, the performance of filtering queries varies with the decreasing size. In *BerlinMOD*, only 2000 attributes are considered for filtering which is a very small sample as compared to 10 million points in

T-drive. On a small dataset, with a larger number of nodes configured, the parallel processing performance decreases. Therefore Query 2 did not perform well in parallel.

Query 3

Where have the vehicles with licenses from Querylicenses1 been at each of the instants from QueryInstants1?

The query is written in sequential as:

```
let SQ_OBACRres003 =
Querylicenses feed {LL} head[10]
loopjoin[dataSCcar_license_btree
dataSCcar exactmatch[.license_LL]]
QueryInstants feed {II} head[10]
Product
projectextend[; license: .license_LL,
Instant: .Instant_II,
Pos: val(.Journey atinstant .Instant_II)]
consume;
```

The query is written in parallel as:

```
let OBACRres003 =
Querylicenses_Top10_Dup_List
hadoopMap["Q3_Result", DLF; .
loopjoin[ para(dataSCcar_license_btree_List)
para(dataSCcar_List)
exactmatch[.license] {LL} ]
para(QueryInstants_Top10_Dup_List)
feed {II} product
projectextend[; license: .license_LL,
Instant: .Instant_II,
Pos: val(.Journey_LL atinstant
.Instant_II)] ]
collect[] consume;
```

As we can observe the *exactmatch* operator is depending on an additional *para* operator in parallel Query 3 which degraded the performance. However the sequential query with no additional operators presents an exceptional performance. It can be inferred that for simple operations such as filtering, exact matching on small data is not effective in parallel mode.

Query 4

What are the pairs of license plate numbers of trucks, which have ever been as close as 10 m or less to each other?

Query 4 of *BerlinMOD* in Parallel Secondo is shown below:

```
let OBACRres006 = dataSCcar_List
hadoopMap[DLF, FALSE; . feed
filter[.Type = "truck"]
extendstream[UTrip: units(.Journey)]
extend[Box: scalerect(bbox(.UTrip),
SCAR_WORLD_X_SCALE, SCAR_WORLD_Y_SCALE,
SCAR_WORLD_T_SCALE)]
projectextendstream[license, Box, UTrip
;Cell: cellnumber(.Box, SCAR_WORLD_GRID_3D) ] ]
dataSCcar_List
hadoopMap[DLF, FALSE; . feed
filter[.Type = "truck"]
extendstream[UTrip: units(.Journey)]
extend[Box: scalerect(bbox(.UTrip),
SCAR_WORLD_X_SCALE, SCAR_WORLD_Y_SCALE,
SCAR_WORLD_T_SCALE)]
projectextendstream[license, Box, UTrip
;Cell: cellnumber(.Box, SCAR_WORLD_GRID_3D) ] ]
hadoopReduce2[Cell, Cell, DLF, PS_SCALE
; . sortby[Cell] {V1} .. sortby[Cell] {V2}
parajoin2[ Cell_V1, Cell_V2; . . .
realJoinMMRTreeVec[Box_V1, Box_V2, 10, 20]
realJoinMMRTreeVec[Box_V1, Box_V2, 10, 20]
filter[(.license_V1 < .license_V2)
and gridintersects(SCAR_WORLD_GRID_3D,
.Box_V1, .Box_V2, .Cell_V1)
and sometimes(distance(.UTrip_V1,
.UTrip_V2) <= 10.0) ]
projectextend[; license1:
.license_V1, license2: .license_V2] ]
sort rdup]
collect[] sort rdup consume;
```

This query provides the optimal performance in the parallel environment with the ideal number of nodes as 6.

Figure 7 gives the overall performance in terms of time with respect to number of nodes. The remaining queries along with the detailed description can be seen in (Duntgen et al. 2009).

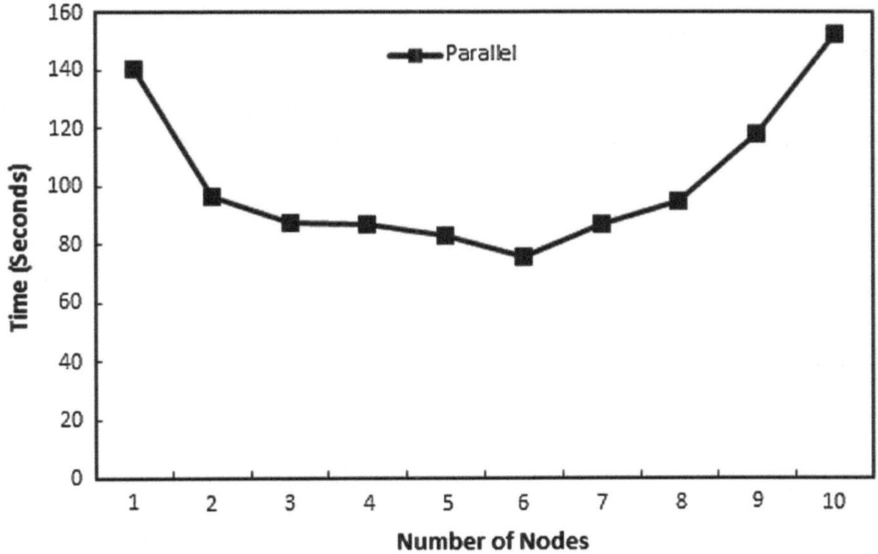

Fig. 7 Computational efficiency of multiple nodes on BerlinMOD

Conclusions and Future Work

This paper describes a series of experiments which were carried out in a distributed spatio-temporal DBMS, Parallel Secondo, to test its efficiency. Parallel Secondo uses the Hadoop framework to share tasks across multiple processing nodes. Different datasets with varying numbers of data points and levels of attribute information were queried using standard spatio-temporal queries in sequential and parallel frameworks of Parallel Secondo. The results highlight that increasing the number of nodes in the distributed system will not always produce efficiency. This is due to the unavoidable overhead of distributing the processing to multiple nodes. The number of data points and the complexity of the spatio-temporal query are the principle factors which determine if parallel processing will be more efficient. The results of the study suggest optimal node numbers for different cases of queries and different volumes of data. This will serve as a guide for researchers wishing to improve the efficiency of spatio-temporal queries on large datasets.

In future, we plan to use *Amdahl's law* in order to check the distribution strategy of Parallel Secondo and the type of query being run. *Amdahl's law* is defined as "a law governing the speedup of using parallel processors on a problem, versus using only one serial processor." In the current study, the default data partitioning strategy of Parallel Secondo is not optimised for all kinds of queries. The users should calculate the probability of each type of query *BerlinMOD* has (e.g. range, point, join etc.) and based on that decide the best partitioning strategy which has the least overall querying cost.

In this paper we tested two very different datasets; in the future we will test more datasets such as Brinkhoff Generator[2] dataset. We will also examine other distributed geospatial big data platforms such as PostGIS,[3] Spatial Hadoop[4] and GIS Tools for Hadoop.[5] This future comparative study will provide a comprehensive analysis of the possible efficiency gains of state-of-the-art distributed spatial data frameworks for spatio-temporal queries.

References

Dean, J., & Ghemawat, S. (2008). Mapreduce: Simplified data processing on large clusters. *Communications of the ACM, 51*(1), 107–113.

Duntgen, C., Behr, T., & Guting, R. H. (2009). Berlinmod: A benchmark for moving object databases. *The VLDB Journal, 18*(6), 1335–1368.

Eldawy, A. (2014). Spatialhadoop: Towards flexible and scalable spatial processing using mapreduce. In *Proceedings of the 2014 SIGMOD PhD Symposium* (pp. 46–50). New York: ACM.

Gao, S., Li, L., Li, W., Janowicz, K., & Zhang, Y. (2014). Constructing gazetteers from volunteered big geo-data based on hadoop. *Computers, Environment and Urban Systems, 61*, 172–186.

Golpayegani, N., & Halem, M. (2009). Cloud computing for satellite data processing on high end compute clusters. In *Cloud Computing, 2009. CLOUD'09. IEEE International Conference on, IEEE* (pp. 88–92).

Guting, R. H., Almeida, V., Ansorge, D., Behr, T., Ding, Z., Hose, T., et al. (2005). Secondo: An extensible dbms platform for research Prototyping and teaching. In *Data Engineering, 2005. ICDE 2005. Proceedings. 21st International Conference on, IEEE* (pp. 1115–1116).

Lee, J.-G., & Kang, M. (2015). Geospatial big data: Challenges and opportunities. *Big Data Research, 2*(2), 74–81.

Li, J., Meng, L., Wang, F. Z., Zhang, W., & Cai, Y. (2014). A map-reduce-enabled SOLAP cube for large-scale remotely sensed data aggregation. *Computers & Geosciences, 70*, 110–119.

Liu, W., Zheng, Y., Chawla, S., Yuan, J., & Xing, X. (2011). Discovering spatio-temporal causal interactions in traffic data streams. In: *Proceedings of the 17th ACM SIGKDD International Conference on Knowledge Discovery and Data Mining* (pp. 1010–1018). New York: ACM.

Lu, J., & Guting, R. H. (2014). Parallel secondo: A practical system for large-scale processing of moving objects. In: *Data Engineering (ICDE), 2014 IEEE 30th International Conference on, IEEE* (pp. 1190–1193).

McArdle, G., Tahir, A., & Bertolotto, M. (2015). Interpreting map usage patterns using geovisual analytics and spatiotemporal clustering. *International Journal of Digital Earth, 8*(8), 599–622.

Murthy, A. C., Douglas, C., Konar, M., O'Malley, O., Radia, S., Agarwal, S., et al. (2011). *Architecture of next generation apache hadoop mapreduce framework*. Apache Jira.

Orakzai, F., Calders, T., & Devogele, T. (2015). Distributed convoy pattern mining. In *ACM SIGSPATIAL 2015* (pp. 4–pages).

Orakzai, F. M. (2014). Trajectory Data Modeling and Processing in HBase. Ph.D. thesis, Technische Universitat Berlin.

---

[2]http://iapg.jade-hs.de/personen/brinkhoff/generator/.

[3]http://postgis.org.

[4]http://spatialhadoop.cs.umn.edu/.

[5]https://esri.github.io/gis-tools-for-hadoop/.

Pelekis, N., Theodoridis, Y., Vosinakis, S., & Panayiotopoulos, T. (2006). Hermes—A framework for location-based data management. In *Advances in Database Technology-EDBT 2006* (pp. 1130–1134). Springer, Berlin.

Tahir, A., McArdle, G., & Bertolotto, M. (2014). A geovisual analytics approach for mouse movement analysis. *International Journal of Data Mining, Modelling and Management, 6*(4), 315–332.

Yu, Y., Zhao, J., Wang, X., Wang, Q., & Zhang, Y. (2015). Cludoop: An efficient distributed density-based clustering for big data using hadoop. *International Journal of Distributed Sensor Networks, 2015,* 1–13.

Yuan, J., Zheng, Y., Zhang, L., Xie, X., & Sun, G. (2011). Where to find my next passenger. In *Proceedings of the 13th international conference on ubiquitous computing* (pp. 109–118). New York: ACM.

Zhu, Y., Zheng, Y., Zhang, L., Santani, D., Xie, X., & Yang, Q. (2012). Inferring taxi status using gps trajectories. *arXiv Preprint* arXiv:1205.4378

Integrated Geo-information Database for Geological Disposal of High-Level Radioactive Waste in China

Peng Wang, Yong-an Zhao, Min Gao, Shu-tao Huang, Ju Wang, Lun Wu and Heng Cai

Introduction

Deep geological disposal is widely considered the most suitable option to deal with high-level radioactive waste (HLW). In China, geological disposal of HLW has entered a critical stage (Wang 2010). Moreover, the development of a geo-information database is an important component in the HLW disposal research and development (R&D) process. In developed countries, research fields related to HLW disposal typically develop and apply information technologies such as data management and data mining. For example, the Nirex Digital Geological Database holds extensive information relating to the Sellafield disposal site in England (Hawkins 2007), and Japan has developed an effective information system for radioactive waste disposal (IAEA CN-1353–7 2005). The Swedish Nuclear Fuel and Waste Management Company began developing the Geo-Tab database for site selection in the 1990s (Eriksson et al. 1992). A site characterization database, i.e. SICADA, which covers multisource research data, has also been developed (Kärnbränslehantering 2000). These two databases have provided a powerful data entity basis for data development and utilization throughout the disposal process. The French National Radioactive Waste Management Agency has developed three major information management systems for the Meuse/Hante Marne underground research laboratory (URL). These systems include a geo database for scientific research data, an SAGD management system for dynamic monitoring of URL data

P. Wang (✉) · Y. Zhao · M. Gao · S. Huang · J. Wang · H. Cai
CNNC Key Laboratory on Geological Disposal of High Level Radioactive Waste Beijing Research Institute of Uranium Geology, 100029 Beijing, People's Republic of China
e-mail: feiyu618@126.com

Y. Zhao · L. Wu
Institute of Remote Sensing and Geographic Information Systems, Peking University, 100871 Beijing, People's Republic of China

C. Zhou et al. (eds.), *Spatial Data Handling in Big Data Era*, Advances in Geographic Information Science,
DOI 10.1007/978-981-10-4424-3_2

and a powerful distributed document management system (Mangeot et al. 2012). This method of organizing related management systems according to different user demands should receive due consideration. Information technology-based research into HLW disposal started late in China, which mainly focused on the practical applications of geographic information systems (GIS) and data management technology in specific fields related to HLW disposal (Li et al. 1998, 2007; Gao et al. 2010; Zhong et al. 2010; Wang et al. 2013, 2014a, b, c). This study will introduce the latest research in geo-information models, integrated geo-information databases and management system development.

Construction of Geo-information Model

An effective data management system for HLW disposal requires development of a geo-information model and construction of a geo-information database. Different types of geo-information data obtained in the previous site selection processes must be considered. To explicitly express data features and connections between different types of data, a geo-information model must be able to handle datasets of various data types such as geographical, geological and geochemical datasets. Therefore, a geo-information model is required prior to developing a geo-information database and management system. When developing a geo-information model, it is extremely important to consider the logical and physical relationships of various data.

Logic Design of Geo-information Model

The main task of logical design is to describe the logical structure of the geo-information database. This task primarily focuses on designing the data structure. At the current stage of HLW disposal, three levels of data features can be obtained and described, i.e. data features associated with a pre-selected area (biosphere), a site (lithosphere) and the rock surrounding the repository. According to ten criteria listed in the *Site selection criteria for an URL for geological disposal of high level radioactive waste in China* (HAD401/06 2013), datasets can be classified into various types such as geological, geographical, hydrogeological and geochemical datasets. Therefore, based on the domains from which existing data are derived and subsequent data expansion, a logical model was established to describe a data entity (i.e. specific to each type of data object) (Fig. 1).

As shown in Fig. 1, considering the characteristics of non-spatial and spatial data, non-spatial data are stored as attribute data and spatial data are organized and stored as map layer objects related to a point, polyline or polygon, which are controlled by metadata. The storage format for spatial data is either well-known binary or well-known text (PostgreSQL 9.2.6 Documentation 2014).

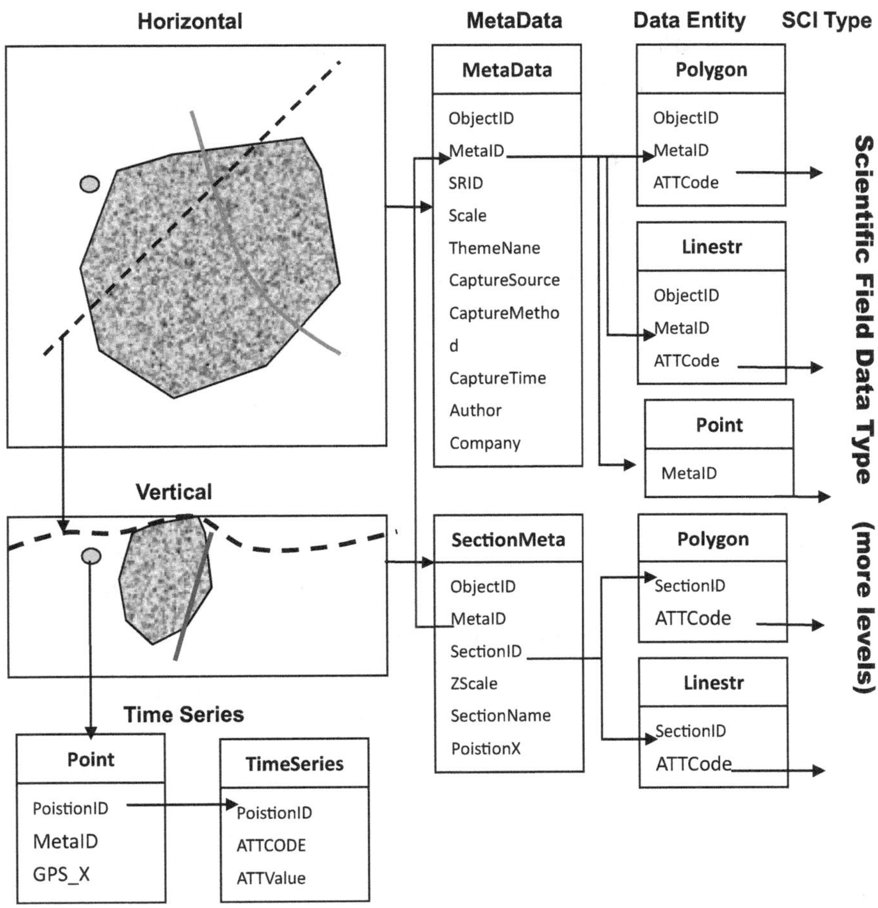

Fig. 1 Logical data model for data associated with geological disposal of HLW

Physical Design of Geo-information Model

Multisource and multidisciplinary thematic data are the main components of a geo-information database for areas pre-selected for HLW disposal. The main content of thematic data and their mutual relations are shown in Table 1. Thematic databases are the primary components of the physical design of a geo-information model. The first design step is to define mutual relations between different datasets and/or data types. Then, different storage methods for different types of data are considered. Finally, the consistency and integrity of the data storage and data expression are achieved.

Table 1 Brief description of data content and mutual relations of geo-information database for pre-selected HLW disposal area

ID	Sub-database	Classification of data entity	Data type	Relationship illustration
1	Metadata database	Metadata information: identification, data quality, spatial reference, content and distribution and the responsible department's contact information	Spatial data, attribute data	Basic descriptive information for thematic data, construction of data dictionary, index reference for all other data
2	Basic geography database	Topography, transportation, pipelines, hydrographic net, geomorphology, vegetation, administrative area and protection zone	Vector data (point, polyline and polygon), attribute data	Basic data such as administrative district, relevant to other data through geometry field
3	Geology database	Rock mass data, characteristics of rock mass, tectonic, fault, stratum, geological boundary, fracture, minerals and alteration, geological section and label	Vector data (point, polyline and polygon), attribute data	Basic geology data, such as fault, lithology, geological boundary, relevant to other data through geometry field
4	Borehole database	Basic borehole information: Engineering geology, geology logging, hydrogeological logging, geophysical logging, hydrologic monitoring, daily drilling records, documentation	Vector data (point, polyline and polygon), attribute data	A series of data obtained around boreholes, relevant to other data through borehole ID and depth fields
5	Remote sensing database	Remote sensing data, target spectral data, image data descriptions, geographical environment, atmospheric environment, measuring method, instrument and equipment, typical spectrum	Raster data (image, photo), attribute data	Relevant to other databases through geometry field

(continued)

Table 1 (continued)

ID	Sub-database	Classification of data entity	Data type	Relationship illustration
6	Hydrology database	Surface water, underground water, geology body, geologic body, hydrological experiment and analytical test	Vector data (point, polyline and polygon), attribute data (test results)	Relevant data of hydrology scientific research field, relevant with other data through geometry field
7	Geophysical database	Airborne geophysical prospecting, geophysical logging, ground physical exploration, interpreted results for physical exploration	Vector data, attribute data and raster data	Borehole geophysical survey can be connected to the Borehole database through the borehole ID, Surface geophysical survey can be connected to the geology database through section ID
8	Geochemistry database	Field test data, indoor sample analysis results, geochemical exploration maps and results	Spatial data (vector and raster data), attribute data	Relevant to sample database through sample ID and geometry field
9	Rock mechanic database	Field test data, laboratory test data, regional survey, digital simulation for test results	Spatial data (vector and raster data), attribute data	Relevant to sample database through sample ID and spatial geometry field
10	Hazardous database	Thematic data such as earthquake, volcano and climate and historical data storage	Spatial data (vector and raster data), attribute data	Thematic data, such as natural hazards, relevant to other databases through spatial geometry field
11	Ecological environment database	Environmental impact assessment data	Spatial data (vector and raster data), attribute data	Relevant to other databases through spatial geometry field
12	Documents database	Achievements reports, scientific reports, domestic and foreign literature	Attribute data, documentation	Relevant to other databases through spatial geometry field
13	Photo database	Scientific results image, thematic images and photos	Vector data, attribute data and raster data	Relevant to other databases through spatial geometry field
14	Sample database	Sample information descriptions, sample locations	Spatial data (vector and raster data), attribute data	Connected to other databases through sample ID and geometry field

Construction of an Integrated Geo-information Database

An integrated geo-information database can be constructed based on the design of the aforementioned geo-information model. First, a powerful database management system (DBMS) should be selected. Considering the unstructured and multisource characteristics of the data, the DBMS should support object-oriented functions such as geometric object abstraction and establishment. To fulfil data storage and retrieval requirements associated with huge amounts of data, the DBMS should also support partition table and partition index technology, parallel data processing technology and distributed database construction technology (Coronel et al. 2011). In this study, the geo-information model and integrated database are both based on PostgreSQL, which is a powerful open-source object-relational database (PostgreSQL 9.2.6 Documentation 2014).

The geo-information database for a pre-selected area (PAGD) has been designed to facilitate the management of a large amount of multidisciplinary research data. The PAGD is classified into sub-databases according to specific professional disciplines. Therefore, there are clear dependency relationships that can be used to establish the hierarchical structure. Constraint conditions, such as major key, unique key, foreign key and default values, are used to correlate information and facilitate data transfer between tables or between spatial and non-spatial data tables. Despite there being some differences between the spatial data and attribute data in the database, the PostgreSQL DBMS can handle the differences easily (PostgreSQL 9.2.6 Documentation 1996–2014). The spatial data can be represented as a geometry column that can be stored and managed in the same way as other data. This will facilitate the realization of the data structure and the organization of data tables.

As shown in Fig. 2, all the data or information related to boreholes can be obtained and organized through the Borehole ID. The borehole spatial position data can be taken as a single geometry column in the BH_General_Info table. Thus, it is easy to correlate and retrieve such data.

Development of Management System for Integrated Geo-information Database

Accomplishment of Metadata Management

In general, metadata describes other data; in particular, metadata can describe a resource object and helps the management, positioning, acquisition and utilization of a data object. Therefore, the integrated geo-information database includes a metadata database that is used for data management, data queries and distribution of

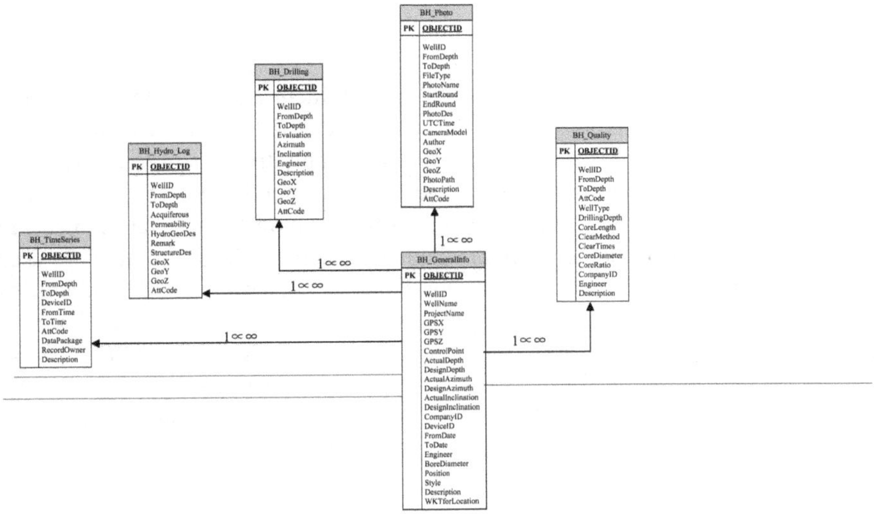

Fig. 2 Examples of data table organization in the Borehole sub-database

Fig. 3 Metadata management interface: ① Function menu area, ② Data directory list area and ③ Metadata display area

all datasets and data features. A robust metadata manager that unifies metadata management was developed to facilitate consistent descriptions and associations. As shown in Fig. 3, data management functions include data preview, additive, maintenance and query functions.

Development of Management System

Development of an appropriate data model and an inclusive, well-structured database are fundamental prerequisites for an efficient data management system. However, the ultimate objective is to retrieve and apply the data. Therefore, given the characteristics of geo-information data and the application requirements, a hybrid C/S and B/S architecture was adopted to accomplish data management. In addition, Open Geospatial Consortium standards and the TCP/IP protocol are used for database management and connectivity. Finally, based on a function module of commercial GIS software, a technological development framework was designed and achieved (Fig. 4). The accomplishment of the framework indicates that the geo-information data model can be developed and realized during the process of secondary development. Thus, a powerful management system can be developed and realized. The main interface of the data management system is shown in Fig. 5.

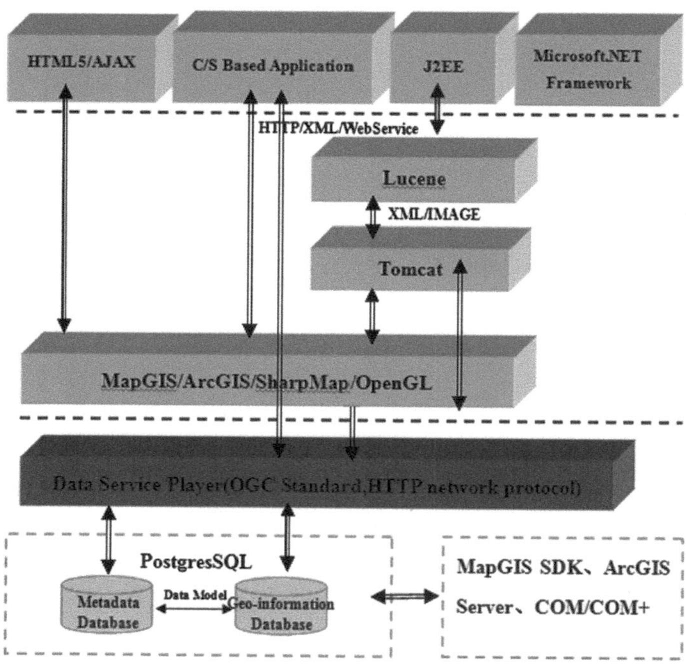

Fig. 4 Technology framework

Fig. 5 Main interface of HLW-GIS management system: ① Function menu area; ② List area for data source; ③ Visual expression area; ④ Data directory and detailed description; and ⑤ Attribute data area* (*The database and management system are both designed for HLW disposal R&D teams in China. Therefore, the system content and user interface are in Chinese.)

Conclusions

Construction of a geo-information data model and development of a data management system are vital steps toward the parts for digitalization construction of an integrated geo-information database system for a pre-selected area. The model and system can also provide technical support for the development of HLW disposal. Some important conclusions are summarized as follows.

(1) Many different types of data are generated during the site selection process for HLW disposal, which are useful for determining the complexity of the geo-information model and database construction. Classification of thematic data is the key to constructing the geo-information model, which should select the most stable and essential attributes as a basis for classification. The integrity and generality of classification results should also be ensured.
(2) An integrated geo-information database was built using the geo-information model. This integrated database will be a powerful foundation for HLW disposal R&D.
(3) A technical framework that facilitates the development and realization of a data management system was developed based on the integrated geo-information database. It directly confirmed the reliability of the geo-information model and the feasibility of secondary development. In addition, the data management

system and the framework are expected to provide solid technical support for future data mining work.

References

Coronel, C., Steven, M., & Peter, R. (2011). *Database systems: Design, implementation and management* (9th ed.), Library of Congress Control Number: 2009936830.

Eriksson, E., Johansson, B. et al. (1992). *GEOTAB. Overview*. SKB, TR92-01.

Gao, M., Huang, S. T., Wang, S. H., & Zhong, X. (2010). Metadata design and editing module development for the geological disposal of high-level radioactive waste. *World Nuclear Geoscience, 25*(4), 37–41. (In Chinese).

HAD401/06, Site selection criteria for an URL for geological disposal of high level radioactive waste in China, 2013. The national nuclear safety administration.

Hawkins, C. (2007). *Geosphere characterisation project-data management strategy*. Tessella Project Number 4998. Nirex, England.

IAEA. (2005). *International Conference on the safety of radioactive waste disposal*. Tokyo, Japan: IAEA CN-1353–7.

Kärnbränslehantering, S. AB. SKB. (2000). *Geoscientific programme for investigation and evaluation of sites for the deep repository* (SKB Technical Report TR-00-20). ISSN 1404-0344.

Li, H. B., Huang, S. T., & Zhao, Y. A. (2007). WebGIS based geo-information system for Beishan disposal repository of high level radioactive waste. *World Nuclear Geoscience, 24*(1), 39–43. (In Chinese).

Li, J., Fan, Ai, Huang, S. T., & Wang, J. (1998). Development of a geoscience database for preselecting China's high level radioactive waste disposal sites. *World Nuclear Geoscience, 14* (2), 107–111. (In Chinese).

Mangeot, A., Tabani, P., Yven, B., & Dewonck, S. (2012). *3D visualization of geo-scientific data for research and development purposes. Clays in natural and engineered barriers for radioactive waste confinement—5*. International meeting. Ref. Number 44067667, Rel. Record 44048818, Publ. INIS Volume 44.

PostgreSQL 9.2.6 Documentation. By The PostgreSQL Global Development Group, Copyright © 1996–2014 The PostgreSQL Global Development Group.@@@

Wang, J. (2010). High level radioactive waste disposal in China: Update 2010. *Journal of Rock Mechanics and Geotechnical Engineering, 2*(1), 1–11.

Wang, P., Li, X. Z., et al. (2013). Geostatistical analysis of fracture density of granite rock in Beishan area Gansu Province based on GIS. *Engineering Geology, 21*(1), 115–122. (In Chinese).

Wang, P., Huang, S. T., Gao, M., Zhao, Y. A., & Wang, S. H. (2014a). Operation environment construction of geological information database for high-level radioactive waste geological disposal. *World Nuclear Geoscience, 31*(1), 299–304. (In Chinese).

Wang, P., Li, X. Z., Wang, J., Zhang, Y. S., et al. (2014b). Research on spatial patterns of fractures in granite rock based on GIS. *Engineering Geology, 22*(6), 1086–1093. (In Chinese).

Wang, P., Huang, S. T., Zhao, Y. A., & Wang, H. B. (2014b). Development of data applications and presentations for geological information database of HLW disposal. *World Nuclear Geoscience, 31*(1), 299–304. (In Chinese).

Zhong, X., Wang, J., Huang, S. T., Wang, S. H., & Gao, M. (2010). Design of geo-metadata in GIS for pre-selected disposal site of high-level radioactive waste. *World Nuclear Geoscience, 27*(4), 219–222. (In Chinese).

Analyzing the Uncertainties of Ground Validation for Remote Sensing Land Cover Mapping in the Era of Big Geographic Data

Bo Sun, Xi Chen and Qiming Zhou

Introduction

Ground validation is a vital process in remote sensing land cover classification. Accuracy assessment provides important information for the users of the classification products, especially for land cover change detection (Olofsson et al. 2013). Ground references or in situ data have been widely accepted in accuracy assessment of remote sensing image classification (Liu and Zhou 2004; Lillesand et al. 2015). However, obtaining ground reference data is often difficult in terms of the number of sample cases and data quality (Foody 2010). In addition, it is almost impossible to get historical ground measurement data. In modern times, the majority of ground reference data come from field survey as well as manual or automatic image interpretation using the original or finer-resolution images.

B. Sun (✉) · X. Chen · Q. Zhou
Center for Geo-spatial Information, Shenzhen Institutes of Advanced Technology, Chinese Academy of Sciences, Shenzhen 518055, People's Republic of China
e-mail: sunbo@siat.ac.cn

X. Chen
e-mail: chenxi@ms.xjb.ac.cn

Q. Zhou
e-mail: qiming@hkbu.edu.hk

X. Chen
Xinjiang Institute of Ecology and Geography, Chinese Academy of Sciences, Urumqi 830011, People's Republic of China

Q. Zhou
Department of Geography, Hong Kong Baptist University, Kowloontong, Hong Kong, People's Republic of China

C. Zhou et al. (eds.), *Spatial Data Handling in Big Data Era*, Advances in Geographic Information Science, DOI 10.1007/978-981-10-4424-3_3

Given the coming era of big geographic data, it is speculated that a huge amount of ground reference data from various sources will become available. For instance, crowdsourcing is a popular way to collect geospatial data at a lower cost by a group of people or community (Heipke 2010).

A widely accepted description of big data characteristics consists of three "V"s, namely "Volume", "Variety", and "Velocity" (Liu et al. 2016). Nowadays, the fourth "V"—Veracity has been highlighted (Li et al. 2016; IBM 2016), suggesting the critical importance of the data quality or uncertainty assessment in the field of spatial data handling.

In most studies on remote sensing image classification, ground reference data is usually regarded as "ground truth" and typically kept unchallenged as correct cases. It should be noted that the reference data technically are just another set of derived outcomes from data processing so that they may also contain errors (Zhou et al. 1998; Foody 2002). It is pointed out that the uncertainties would exist in inaccurate locations, wrongly classified labels, misinterpretation of the meaning of labels, and so on (Liu and Zhou 2004).

This study aims to analyze the uncertainties of ground validation for remote sensing classification based on a large amount of reference data which may or may not be reliable. In addition, the impact of the reference data error on the overall accuracy assessment result is also analyzed.

Methodology

Study Area and Test Data

A case study has been undertaken to evaluate the accuracy of a land cover classification in Central Asia. Central Asia normally refers to the central hinterland of the Eurasian Continent. The land cover classification covers five nations, namely, Kazakhstan, Tajikistan, Turkmenistan, Uzbekistan and Kyrgyzstan. The total mapping area is around 4 million km^2. The environment here is generally dry, with arid and semi-arid ecosystems. Most areas are unfrequented and covered by desert or bare land.

The 2010 land cover classification of Central Asia was utilized in the test. The data set was produced by Xinjiang Institute of Ecology and Geography (XIEG) of Chinese Academy of Sciences (CAS) by interpreting Landsat series images with a spatial resolution of 30 m. An object-orient classification method with manual interpretation was adopted for land cover classification. Seven first-level land cover types were identified, namely, cultivated land (C), forest (F), grassland (G), artificial surface (A), transportation (T), water body (W), and bare land (B).

Sampling Scheme

Since the study region covers a huge areal extent, and many places are arduous to get assess to, collection of ground reference data based on field investigation would be unpractical because of the extremely high cost in financial resource and time. Besides, historical ground data are either too difficult to obtain, or, in most cases, do not exist.

To balance the cost and the efficiency of sampling, a dual-model sampling scheme was adopted in the study (Table 1). Similar to many other studies, reference data based on manual interpretation of finer-resolution images was utilized to assess the accuracy of the classification. In addition, limited ground reference data were collected through field investigation. The ground-based reference data are assumed more reliable, thus they were utilized to evaluate the above reference data sets derived by the manual interpretation.

Normally, a random sampling scheme with a large sample size allows more accurate and stable statistics (Stehman and Czaplewski 1998; Tsendbazar et al. 2015). Given the large diversity in spatial patterns shown by different land cover types, a stratified random sampling scheme was adopted for the image-based sampling (model 1). According to patch density or complexity of the land cover classification, 6301 sampling points were deployed. For the areas with complex land cover features, additional high-density sampling was employed. 300 ZY-3 (China's Earth resource satellite) multi-spectral images at a high spatial resolution of 6 m were used for this. The acquisition period of the images was in 2012, not ideally synchronized with the 2010 test data sets but close enough for classification evaluation. For each ZY-3 image, 70 points were randomly sampled, giving totally more than 27,000 samples (Fig. 1).

340 ground reference samples were collected through field investigation (model 2). Field investigations were conducted in collaboration with the agencies of the Central Asian countries during 2012–2014. Figure 2 illustrates the distribution of the ground samples.

The land cover type at each sampling point was identified by manual interpretation on a higher-resolution image such as those from Google Earth and China's

Table 1 The dual-model sampling scheme

	Model 1	Model 2
Sample source	Fine-resolution image interpretation	Ground measurement from field investigation
Amount (points)	26,282	315
Distribution	Stratified randomly	Designated
Trust degree	Highly trusted	Close to truth

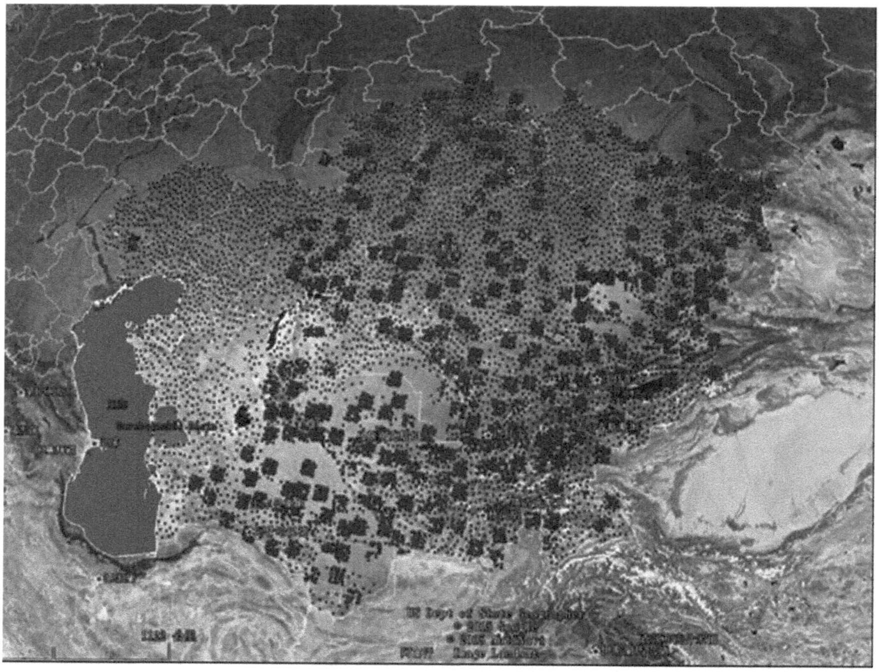

Fig. 1 Distribution of reference samples for image interpretation over the study area (Google Earth imagery as background)

Earth resource satellite (ZY-3) images. The image interpretation results were cross-validated by three well-trained image interpreters to ensure maximum confidence of the land cover type identification. A land cover type identification was accepted with an agreement of at least two interpreters. The samples without sufficient agreement would be dropped. After cross-validation, more than 26 thousand references based on image interpretation were used in accuracy assessment of land cover classification. The 315 ground references obtained by field investigation were utilized to evaluate the reference data sets by manual interpretation.

Reliability of Ground References

The correlations of three-time measurements were calculated to test the consistency among the three image interpreters. In addition, the consistency of ground references and corresponding samples taken by the manual interpretation was tested. Thus, the reliability has been analyzed on the reference data set derived by the manual interpretation.

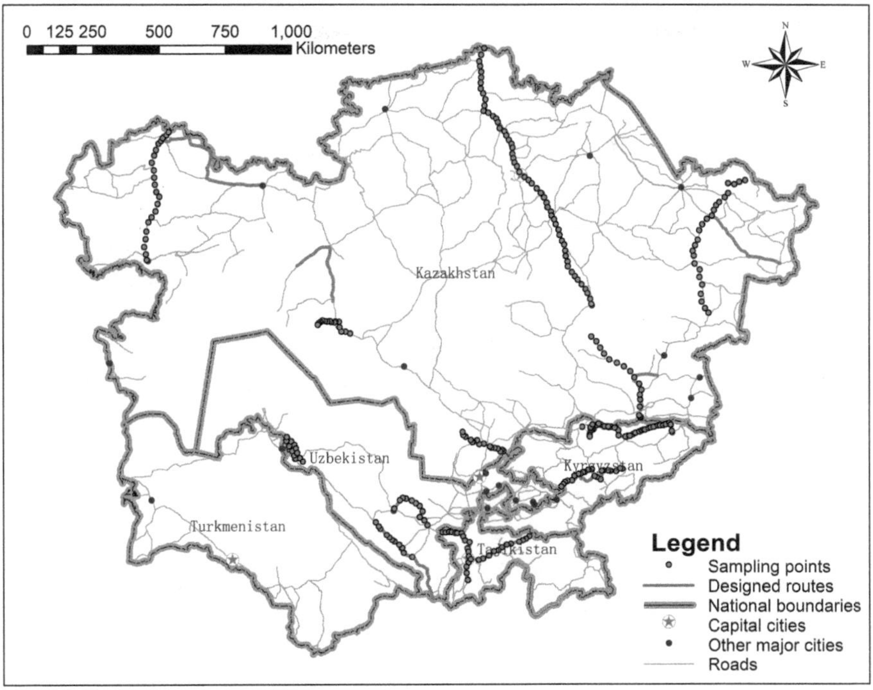

Fig. 2 Distribution of the in situ measurements

Accuracy Assessment

Confusion matrix and *Kappa* coefficient were employed for assessing the accuracies of both the reference data and land cover map. The accuracy measurements include the user's and producer's accuracies for each class. *Kappa* coefficient is calculated based on the error matrix using the following equation (Lillesand et al. 2015).

$$Kappa = \frac{N\sum_{i=1}^{r} x_{ii} - \sum_{i=1}^{r}(x_{i+} \cdot x_{+i})}{N^2 - \sum_{i=1}^{r}(x_{i+} \cdot x_{+i})} \tag{1}$$

where

r = number of rows in error matrix
x_{ii} = number of observations in row i and column i
x_{i+} = total of observations in row i
x_{+i} = total of observations in column i
N = total number of observations included in matrix

Results and Analysis

Overall Accuracy of the Land Cover Classification

Based on the cross-validated reference data set of over 26,000 samples, the 2010 land cover classification shows an overall accuracy of 63% with *Kappa* = 0.44. As for specific land cover types, bare land and cultivated land have the highest producer's (93%) and user's accuracies (86%), respectively, while grassland and forest have lower producer's (32%) and user's accuracies (30%), respectively. Besides, transportation land is almost impossible to be classified correctly at the 30-m scale. The majority of classification uncertainties come from the confusion between grassland and bare land as well as the confusion between forest and grassland.

Consistency of Reference Data

From correlation analysis result, the three-time interpretation results are significantly correlated with the final interpretation result (0.65, 0.80, and 0.85, respectively with $p < 0.05$). Compared with in situ observation data, reference data set derived by the manual interpretation shows an overall accuracy of 50% (Table 2).

As for specific types, cultivated land has a relative high image interpretation accuracy of 63%, while the others' accuracies are around 50%, except for transportation land (18%) and water body (17%). Reliabilities of manual interpretation for different land cover types in the study region are shown in Fig. 3.

Table 2 Confusion matrix of image interpretation based references *vs.* in situ measurements

	In situ measurement[a]								
Image interpretation		C	F	G	A	T	W	B	tot
	C	40	0	8	6	0	1	9	64
	F	3	16	4	6	0	5	3	37
	G	9	5	46	2	4	2	18	86
	A	4	3	4	13	2	0	0	26
	T	2	2	1	3	2	0	1	11
	W	2	0	1	1	0	1	1	6
	B	9	1	23	9	0	3	40	85
	tot	69	27	87	40	8	12	72	
	Overall accuracy = 158/315 = 50.2% *Kappa* coefficient = 0.374								

[a]Land cover types: *C* cultivated land; *F* forest; *G* grassland; *A* artificial surface; *T* transportation; *W* water body; *B* bare land

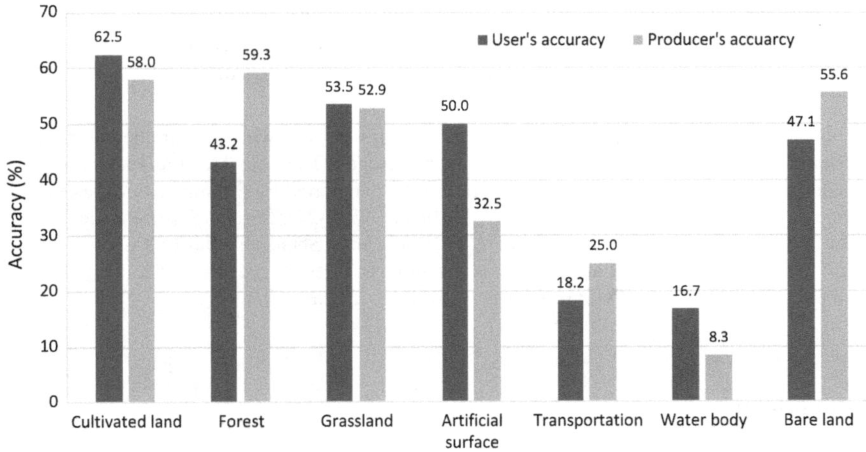

Fig. 3 Reliabilities of manual image interpretation by land cover type

Discussion and Conclusions

Cross-validation result indicates that independent manual interpretation by three persons is consistent with each other. However, from the reliability analysis result, the manual interpretation of finer-resolution images cannot be proved as a high-quality source of ground reference data. Many studies have discussed the impact of errors presented in the reference data on accuracy assessment of land cover classification from the view of error propagation or accumulation (Goodchild et al. 1992; Foody 2010). Given the confusion between grassland and bare land yielded major errors of image interpretation in this study, insufficiently reliable grassland (or bare land) references would cause misleading accuracies not only in grassland classification but bare land classification.

To analyze the uncertainties, scale issue is very common in remote sensing image classification and accuracy assessment. For example, road can be correctly recognized in field survey but misclassified by image classification or interpretation at a scale of 30-m spatial resolution.

Besides, seasonal effect may have influences on data accuracies. Especially in the arid region, an ephemeral stream in arid areas could be identified as "water body" in wet seasons but "bare land" in dry seasons. This might be the reason why water body demonstrated a lower identification accuracy in the study. Moreover, uncertainties may come from the change itself, such as changes from grassland to bare land, since the date of field investigation is different from that of image being acquired to a certain extent.

This study has demonstrated the uncertainties in ground reference data. For the most common source of ground reference data, manual interpretation from Google Earth or other higher-resolution images did not show a high reliability in comparison with ground measurement. Image interpreter's experience is important in this situation. Given the existence of errors in ground reference data, the accuracy of land cover classification should be in a range. A new evaluation system is needed for more reliable accuracy assessment on image classification.

Acknowledgements The authors would like to thank the persons who assist in the interpretation of ground reference data based on high-resolution images. They are Ms. Liu Ping, Ms. Chen Huijuan, Ms. Yi Lin, and Mr. Li Jilin. The research is supported by the International Science & Technology Cooperation Program of China (2010DFA92720-24), Natural Science Foundation of China (NSFC) General Research Grant (41471340) and Shenzhen Basic Research Project (JCYJ20150630114942260). Land cover classification products are provided by Xinjiang Institute of Ecology and Geography (XIEG), Chinese Academy of Sciences (CAS).

References

Foody, G. M. (2002). Status of land cover classification accuracy assessment. *Remote Sensing of Environment, 80,* 185–201.

Foody, G. M. (2010). Assessing the accuracy of land cover change with imperfect ground reference data. *Remote Sensing of Environment, 114,* 2271–2285.

Goodchild, M. F., Sun, G., & Yang, S. (1992). Development and test of an error model for categorical data. *International Journal of Geographical Information Systems, 6*(2), 87–104.

Heipke, C. (2010). Crowdsourcing geospatial data. *ISPRS Journal of Photogrammetry and Remote Sensing, 65*(6), 550–557.

IBM. (2016). *The four V's of big data. Retrieved from IBM Big Data & Analytics Hub website.* http://www.ibmbigdatahub.com/infographic/four-vs-big-data. Available on June 24, 2016.

Li, S., Dragicevic, S., Castro, F. A., Sester, M., Winter, S., Coltekin, A., et al. (2016). Geospatial big data handling theory and methods: A review and research challenges. *Journal of Photogrammetry and Remote Sensing, 115,* 119–133.

Liu, H., & Zhou, Q. (2004). Accuracy analysis of remote sensing change detection by rule-based rationality evaluation with post-classification comparison. *International Journal of Remote Sensing, 25,* 1037–1050.

Liu, J., Li, J., Li, W., & Wu, J. (2016). Rethinking big data: A review on the data quality and usage issues. *ISPRS Journal of Photogrammetry and Remote Sensing, 115,* 134–142.

Lillesand, T. M., Kiefer, R. W., & Chipman, J. W. (2015). *Remote sensing and image interpretation* (7th ed.). New Jersey: Wiley.

Olofsson, P., Foody, G. M., Stehman, S. V., & Woodcock, C. E. (2013). Making better use of accuracy data in land change studies: Estimating accuracy and area and quantifying uncertainty using stratified estimation. *Remote Sensing of Environment, 129,* 122–131.

Stehman, S. V., & Czaplewski, R. L. (1998). Design and analysis for thematic map accuracy assessment: Fundamental principles. *Remote Sensing of Environment, 64,* 331–344.

Tsendbazar, N. E., de Bruin, S., & Herold, M. (2015). Assessing global land cover reference datasets for different user communities. *ISPRS Journal of Photogrammetry and Remote Sensing, 103,* 93–114.

Zhou, Q., Robson, M., & Pilesjo, P. (1998). On the ground estimation of vegetation cover in Australian rangelands. *International Journal of Remote Sensing, 19,* 1815–1820.

Error in Spatial Ecology (PVM)

Brian Lees

Introduction

Uncertainty and error are unavoidable in spatial ecology mapping just as in other geographical applications. There are several common approaches, mapping polygons directly on air photographs, classification of digital remote sensed data or, often the most accurate, the spatial extension of good point data obtained by field survey (Predictive Vegetation Mapping or PVM). The focus of this paper is PVM. Huang and Lees (2004, 2005, 2007) looked at two main types of error in the sorts of models used to spatially extend good, point, vegetation data, inherent and operational. Inherent error is error resulting from inaccurate input data, while operational error is the error which results from the modelling itself. All spatial data, whether they are paper maps or digital layers, in vector or raster format, having categorical or numerical values, contain errors to some extent. This is due to not only instrument and human errors, but also the age of the data and the inherent complexity of the real world. Models tend to reduce the complexity of the real world using approximations and the cost of these approximations is increased error, uncertainty and less precision. Errors in input data can be transmitted through the modelling process and become manifest in the final products. Different models transmit error in different ways. The exact error propagation modes for most models are unclear and need careful analysis (Huang and Lees 2005). Van Niel et al. (2004) did this for error in elevation data, a common input to PVM, and Huang (Huang and Lees 2004) did this for PVM more generally. Huang and Lees (2007) and Hagen-Zanker et al. (2005) went further and looked at the effects of learning sample location and class error.

B. Lees (✉)
School of Physical Environmental and Mathematical Sciences,
University of New South Wales Canberra, ACT, Australia
e-mail: Brian.Lees@adfa.edu.au

C. Zhou et al. (eds.), *Spatial Data Handling in Big Data Era*,
Advances in Geographic Information Science,
DOI 10.1007/978-981-10-4424-3_4

Once we properly identify the types and sources of error, track the error transmission through the modelling process, measure and quantify the magnitude of the error with suitable error indices, and visualise the error measurements for the users, sensible strategies need to be implemented to reduce or manage the error if possible. Common sources of error in predictive vegetation modelling are DTM inaccuracy, which flow through to slope, wetness estimates and aspect/insolation calculations, and soil/lithology data. Given the problems of circularity with soil mapping which is now often based on a model integrating vegetation, geology and slope data (McBratney et al. 2003), vegetation modellers often resort to geology maps rather than soil maps. However, geology maps are rarely produced at fine scale and often the classes depicted relate to the age of a unit rather than its varied lithology. These issues have been dealt with elsewhere.

This paper looks at one of the common sources of error in predictive vegetation mapping, the actual style of mapping, although the principles are more generally applicable to other forms of natural resource mapping. The style of mapping commonly used for vegetation is the area-class map (Mark and Csillag 1989; Bunge 1966). This is a form of thematic map often described as a choropleth map, but Mark and Csillag (1989) point out that this usage is incorrect and that the term 'choropleth map' should be restricted to situations where the bounding polygons are defined by other than the phenomena being mapped. For example, the administrative boundaries used by McHarg and Mumford (1969) for forest type polygons in his Potomac River Basin Study are correctly choropleth maps but give a quite misleading impression of the land cover around Washington because they are inappropriate for the phenomena being mapped and suffer badly, as a result, from the Modifiable Areal Unit Problem.

Background

In 1977 David Sinton wrote a seminal paper on 'the inherent structure of information as a constraint to analysis'. He used mapped thematic data of natural resources as a case study. Thematic mapping dates back to the early 17th Century, well before computer databases were available. However, thematic mapping of natural resource information of the type discussed by Sinton is properly area-class mapping although he didn't use that term. In these types of map the structure is of discrete polygons with the boundaries between categories or classes of the theme. These occur over contiguous regions of geographic space. The polygon structure can be seen as enclosing an Entity but Goodchild et al. (2009) suggest it can also be seen as is a Nominal Field. The structure both reflects the way we think and, to some extent, how we go about collecting data.

Sinton made the point that certain types of information are collected in such a way that there are steps which include generalisation inherent in the collection process. This generalisation of the information collected limits the types of analysis which can be carried out later.

This sounds like an obscure and abstract point, but it is not. For example, during the 1980's the Australian Government decided, as a result of considerable conflict between those supporting the logging of native forests and those opposing this on environmental grounds, to have a National Forest Inventory. This was intended to produce a master data set to avoid the conflict which had arisen between datasets produced by the competing interests. The first pass at joining up forest maps across state borders revealed that there were serious inconsistencies in the ways each state, and in some cases region, defined their forests. This should not have been a surprise as similar problems had been observed when attempts have been made to mosaic soil maps across administrative boundaries. Solving the problem by resorting to the original observations to recode the data proved impossible as they had not been retained.

Indeed, the practice in many field sciences was to observe, classify then record. In many cases the detail of the original observation only existed in field notebooks, which were not archived. In one legacy vegetation map close to home, of the Australian Capital Territory, it was found that the map was a mosaic based on three surveys, each by a different group, each at a different time, and each at a different nominal scale. The apparent homogeneity of one part and the apparent heterogeneity of another part were artifacts of this mosaicking. The map was, as a result, extremely misleading.

The situation described above is common. It is extremely costly to collect new data and the temptation is to make do with what is to hand. So, it is important to properly understand the process by which that data has been generated and stored.

Descriptions, Depictions, or Diagrams

The area-class map is a type of diagram. Maps and images, more generally, fall into this class of media. Often field scientists use a mixture of descriptive writing, symbolic equations and diagrams to communicate their ideas. These media appear to be processed in our brains through different pathways. The distinction between the way we believe these are processed are between model-based theories (including graphical ones), and sententially based theories. These appear to be dealt with by different parts of our brain. The issue has been presented as a choice between reasoning being semantic or syntactic respectively (Engelhardt 2002). In terms of syntax, written text and symbolic equations are essentially linear, while diagrams need not be. This suggests that the common practice of supporting a piece of difficult text with a graph or diagram works because we are delivering the message to our brain through two different pathways.

We can discriminate between the way we understand descriptions (written text and symbolic equations) and diagrams (Lemon and Pratt 1999). Most descriptions are Fregean (Peregrin 2000) where the symbol bears no resemblance to the thing it represents in terms of its properties. But, with analogical representations where relationships are explicit, there is such a resemblance. This is the first key difference between written text and symbolic equations (Fregean), and diagrams (analogical). The second key difference is that there is a sense in which text (and equations) are a sequential set of symbols. Text (and equations) rely for much of their meaning on this sequence. Usually what is written later in a sentence assumes that the reader knows what has been written beforehand. Because written text is processed by people in a sequential fashion it is hard for humans to store all the things that we have just read in our working memory, especially when the number of items appearing in the text is large. Diagrams can avoid this problem. However, sentential descriptions may allow a more accurate and precise description of a phenomenon than diagrams.

Only those diagrams in the bottom half of Table 1 can deal with spatial data spatially. This distinguishes tables and charts from diagrams of the sort we are concerned with. Why is a sentential representation not as effective as a diagram in helping us to solve a problem? (Larkin and Simon 1987);

- Diagrams can group together all the information that is used together. This effective use of display space is an important attribute as it thus avoids lengthy searches for the elements needed to make a problem-solving inference.
- Diagrams avoiding the need to match symbolic labels by using location to group information about a single element.
- Diagrams automatically support a large number of perceptual differences, which are extremely easy for humans to understand. In addition to the relationships in space, the meanings of relative size and colour are all easily understood by humans.

Diagrams appear to operate best when cognitive reasoning, which must extract the structural information from the sentential data by laborious comparisons and computations, is supported by a visual representation from which the user can

Table 1 Types of diagram (modified after Lemon and Engelhardt in TwD 1997)

Type of diagram	Characteristics	Treatment of spatial data is
Columns/list	Partitioning in one dimension	Aspatial
Table/matrix	Partitioning in two dimensions	Aspatial
Time line	Metric in one dimension	Aspatial
Two-axis chart	Metric in the Cartesian plane	Aspatial
Bar chart	Partitioning in one dimension combined with Metric in one dimension	Aspatial
Picture or image	Realistic; planar metric combined with planar topology	Spatial
Map	Symbolic; planar metric combined with planar topology	Spatial
Network diagram	Planar topology	Spatial
3D/CAD graphics	Metric and topology in Cartesian 3-space	Spatial

easily perceive the structure of the data. The use of graphs, tables and bar charts in the scientific literature is the best example of this phenomena. Maps, on the other hand, are quite often 'stand-alone' with the only text accompanying them being peripheral to the image. Some types of map rely on only the users' familiarity with the style of representation to carry their message. The area-class map has been successful because the message that it carries is easy to comprehend.

Collecting Data

Sinton (1977) breaks down the process of creating useful data for area-class mapping from initial observation. He argues that one must record the Theme, the Location and the Time of the observation. At least three types of generalisation take place;

- Aggregation—the transformation of the structure from point to polygon usually involves a loss of detail in Spatial Location of the original observation.
- Classification—observations are categorized with other like observations. In practice 'like' becomes 'similar' and the breadth of what 'similar' means is an important source of error.
- Induction—where a series of samples are generalised to include locations assumed to have the same characteristics. Thus aggregating places with similar (but not the same) characteristics as the places sampled into Classes.

Sinton (1977) defines this more precisely (in the case of vegetation mapping).

- One of the Attributes of the data (Time) is held constant.
- A second Attribute (Theme) is permitted to vary in a controlled way.
- The third Attribute (Location) is measured for its variation within the second (controlled) Attribute.

In thematic mapping, time is rarely not held constant. For some themes, such as elevation, soils or geology, this is rarely a problem. But for other more dynamic themes, such as vegetation, land use, land cover or population, the freezing of the theme at the time of data collection does present a major source of error. Allowing the theme to vary 'in a controlled way' runs the risk of falling foul of the Modifiable Area Unit Problem (Openshaw 1983). In this, poorly represented classes tend to be incorporated into larger classes, and so disappear. Also in many cases a continuum of change is represented as a series of classes, which are essentially overlapping gaussians. This, of course, leads to the creation of errors of omission and commission in the theme. Allowing the location to vary from a point to a polygon leads to similar errors. Generalisation can be seen as inducing error with regard to the original observations.

So, the form of an area-class map certainly aids perception and processing of the information presented. But this is at the cost of considerable generalization.

Changing the Data Model

In a previous paper (Lees 1996b) we investigated a method of representing vegetation data which avoided many of the errors which inherent in the use of the area class map. In this earlier paper we investigated how the processing of data to suit this format perpetuates the use of an inappropriate data model and places an upper limit on the accuracy of spatial extension of point data by most predictive modelling techniques.

A number of projects based on the Kioloa data set to develop and test new predictive modelling tools were set up to use standard (forest) industry data as input, and forest types as output (Moore et al. 1991; Lees and Ritman 1991; Fitzgerald and Lees 1993, 1994, 1996). Using the same modelling methods, learning sample and independent variables, but not classifying the learning samples into communities or forest types beforehand, it was shown to be possible to achieve a significant improvement in predictive accuracy over area-class mapping. If one deals with the point observations of vegetation without classifying them beforehand, then another form of visualisation can be used. Because we are dealing with digital information, we are no longer constrained to produce a single 'map' as the data storage medium and visualisation medium. We can now store and retrieve a considerable amount of information about any point, either in geographic, environmental or spectral space. This is a database-oriented approach. If we retain geographic space as the most convenient operational data space for users of predictive modelling, we are also selecting the domain with least database complexity where each point exists in only one location in each of the other domains (Lees 1994; 1996a, b; Aspinall and Lees 1994). We can then model the spatial extension of each relevant attribute of each entity observed in the ground truth plots.

The Database Oriented Technique

In Lees (1996b) the original observations were recoded as fuzzy membership of the canopy, species by species. A simple Back Propagation neural net was set up for each species. In order to provide an ongoing comparison of methods, the input layer was the same as that described in Fitzgerald and Lees (1993, 1994), and used the same datasets as Lees and Ritman (1991). A 9/10/10 structure was used with one hidden layer of ten nodes and an output layer of ten nodes. No spatial or temporal context was used. Each output node represented a range of fuzzy memberships (0–0.1; 0.1–0.2 and so on). The highest number in the output range was taken to indicate the membership of that cell and the whole output range for each cell was treated as a distribution and the probability of the membership was calculated. So, for each species, it was possible to estimate its fuzzy membership of the canopy and the probability of that estimate. For the species chosen as an example, Eucalyptus maculata, and using only the fuzzy memberships, the RMS error was calculated to be 0.2324. Calculating a distribution based on the ten output nodes is, of course, not optimal, but it seemed more sensible than setting the network up with, say, a 9/10/30 structure.

The constraints we placed on our earlier work to make it match existing user expectations resulted in an upper limit to predictive accuracy, for all the models tested, of about 65%. Those themes which were more appropriately represented by this data structure (land/sea discrimination) could be predicted with up to 99% accuracy. The later work (Lees 1996b) which expressed species distribution as a fuzzy membership of the tallest stratum (in a forest), midstratum or understorey, and lower stratum or ground layer, allows the prediction of a series of data layers which represent surfaces of spatially varying fuzzy memberships appeared to offer a significant gain in accuracy. Further, the use of a simple neural net configuration enables both fuzzy membership and the probability of this membership to be estimated. This made it possible to track error through subsequent uses of the modelled estimates. Methods of comparing the two methods have subsequently been investigated and this paper presents some of these findings.

Just How Bad Is an Area-Class Map of Forest Types?

The method used in Lees (1996b) required the information to be stored in some form of a database and no integration of species distributions was suggested. For use in the field this meant that a small computer system would have to be utilised. A fieldworker wishing information about a point would have to scroll through, in the case of the Kioloa data set, the distributions of forty-one tree species, ninety-four shrubs and one hundred and eight understorey species. Together with an error map for each, a database of 486 coverages would be needed for the Kioloa area. It would certainly be possible to query the database to return the information relating to a point but the overview of the immediate local area would still be elusive.

The increase in accuracy gained is seriously offset by the indigestible nature of the output. For real, practical, use some form of synthetic product is needed. This inevitably leads back to a classification of the data into communities, forest types, and so on. Depiction of this sort of synthesis naturally leads to an area-class map. It was argued in Lees (1996b) that it was easy to show that this is a major source of error but there is no insight into what might be the most appropriate replacement. In this exercise we have examined whether this claim was as accurate as thought. Just how bad is an area-class map of forest types for the normal user of these products?

To evaluate this question Allison (1998) decided to move away from the simple measure of 'overall accuracy' and investigate other measures. The calculation of 'overall accuracy', or 'proportion of samples correctly classified', used in many of our earlier papers, ignores any difference between the accuracy of individual classes. She used the Kioloa data set and one of our early models based on it (Lees and Ritman 1991) for the evaluation. This model was one of the earliest we developed, and one of the least accurate. The data set used covers an area of coastal plain and beach which runs back into a sandstone ridge with elevations ranging from sea level to 285 m. Land cover varies from wet sclerophyll rainforest, highly disturbed dry and wet sclerophyll forests with a complex fire history parts of which have been logged in the past, through heath to cleared grassland. There are small pockets of rainforest with palms. PVM models based on this dataset using the ground data pre-classed into forest types typically give predictions of 'Sea' at better than 95%, 'Paddock' at better than 80% and the other forest types at accuracies of less than 65%.

The 'Overall Accuracy' calculation of the Lees and Ritman (1991) model gave only a moderate result of 47.64%, based on the error matrix (Table 2), 'User's Accuracy' and 'Producer's Accuracy' calculations (Table 3) and errors of omission and commission (Table 4) were calculated for each class. These quite effectively demonstrate the problem of treating the continuum of change within this type of forest as a set of overlapping gaussians, or classes. Classes which were clearly separable, such as 'Sea', were predicted with low values of errors of omission or commission and high values of both Producer's and User's Accuracy. The difficult ecotonal 'classes' such E.botryoides forest and Lower Slope Wet forest, which appear to be separable in geographic space, are not separable in environmental data space and have very large errors associated with them (Lees 1996b). Whilst it is tempting to interpret this as meaning that better results are impossible using classed data, it is worth examining the degree of error involved in these measures.

Forest maps are specifically targeted at a particular type of user, the forest manager. Traditionally, the major concern of such a user was to accurately predict the timber volume in a forest coupe. Multi-use forests have extended the range of concerns foresters must address, but it is possible to look at each use and categorise the level of unsuitability of the map for a particular purpose. In uses where species composition is important, one could generate a table such as this (Table 5).

Table 2 The error matrix for Lees and Ritman (1991)

	Reference data									Total
	1	2	3	4	5	6	7	8	9	
Predicted class										
1	167	15	15	35	35	4	2	7	0	280
2	13	3	1	7	8	2	2	8	0	44
3	13	1	1	4	1	1	1	1	0	23
4	32	20	14	126	43	73	60	45	8	421
5	52	18	8	46	70	6	6	17	2	225
6	6	0	2	17	5	9	3	0	0	42
7	22	8	10	16	19	2	12	18	3	110
8	1	1	1	0	0	0	1	48	1	53
9	0	0	0	0	0	0	0	2	259	261
Total	306	66	52	251	181	97	87	146	273	1495

Table 3 User's and producer's accuracy (Allison 1998)

Vegetation class	User's accuracy (%)	Producer's accuracy (%)
1 Dry Sclerophyll forest	59.64	54.58
2 E.botryoides forest	6.82	4.55
3 Lower slope wet forest	4.35	1.92
4 Wet (E.maculata) forest	29.93	50.20
5 Dry (E.maculata) forest	31.11	38.67
6 Rainforest ecotone	21.43	9.28
7 Rainforest	10.91	13.79
8 Paddocks and cleared	90.57	32.88
9 Sea	99.23	94.87

Table 4 Errors of omission and commission (Allison 1998)

Vegetation class	Error of omission (%)	Error of commission (%)
1 Dry Sclerophyll forest	45.42	40.36
2 E.botryoides forest	95.45	93.18
3 Lower slope wet forest	98.08	95.65
4 Wet (E.maculata) forest	49.80	70.07
5 Dry (E.maculata) forest	61.33	68.89
6 Rainforest ecotone	90.72	78.57
7 Rainforest	86.21	89.09
8 Paddocks and cleared	67.12	9.43
9 Sea	5.13	0.77

Table 5 Levels of disagreement (Allison 1998)

Level 0	No disagreement	Predicted vegetation type matches actual vegetation type exactly
Level 1	Low level disagreement	Predicted vegetation type incorrectly assigned, but both predicted and actual vegetation types contain at least one identical main indicator species
Level 2	Moderate disagreement	Predicted vegetation type main species occurs as associated species or vice versa
Level 3	High level disagreement	Only predicted associated species occurs (as associated species)
Level 4	Total disagreement	No predicted main or associated species found

It can be seen that disagreement at both levels 1 and 2 is not likely to be very significant to the sort of user described above, the prediction is 'fairly close' to the actuality. Level 4 disagreement, however, is simply wrong. Here the prediction is completely at odds with the actuality. Disagreement at level 3 is harder to categorise, it may be as serious as level 4 for some users. But a ranking of the levels of disagreement such as this allows us to categorise the misclassifications in Lees and Ritman (1991) as shown in Table 6.

Table 6 Number of samples at each level of disagreement (Allison 1998)

Level of disagreement	Number of samples	Percentage of total
0	695	47.64
1	190	13.02
2	312	21.39
3	122	8.36
4	140	9.56

We can then go on to quantify the errors by looking at our test sample (Table 6) or even plot these as a map of levels of disagreement. From Table 6 one could suggest that, for a vegetation scientist checking the model, the accuracy would certainly be only 47.64%, but for a forest manager the accuracy would, in practical terms, be between 60.66 and 82.05% as the predicted species does, in fact, occur at level 2 disagreement. This suggests that it can still be a usable product at these levels. Importantly, the latter level of accuracy is slightly higher than that achieved by Lees (1996b).

Conclusion

All of this suggests that some our efforts to identify error and to reduce it have, if not been misguided, ignored the practicalities of field use. Even with all the miniaturized electronic tools available for field use, the database model is too cumbersome a vehicle for easy comprehension of all the necessary information. The traditional form of delivery, and use, of vegetation and soils information, the area-class map, may indeed perpetuate the use of an inappropriate data model and place an upper limit on the accuracy of spatial extension of point data by most predictive models, but it remains a practically useful form of delivering spatial information for most users. Clearly a combination of both methods, that described by Lees (1996b) and the synthesized area-class map, would allow access to the advantages of both. The former is clearly best suited to the office system, and the latter to the field.

References

Allison, B. (1998). *The assessment of the accuracy of a vegetation map produced using a GIS.* Unpublished BSc(REM) honours thesis, Department of Geography, Australian National University, Canberra, p. 133.

Aspinall, R. J., & Lees, B. G. (1994). *Sampling and analysis of spatial environmental data. Advances in GIS research* (pp. 1086–1098). Southampton: Taylor and Francis.

Bunge, W. (1966). *Theoretical geography.* Royal University of Lund, Department of Geography; Gleerup.

Engelhardt, J. V. (2002). The language of graphics: a framework for the analysis of syntax and meaning. Universiteit van Amsterdam. Institute for Logic. Language and Computation, pp. 193.

Fitzgerald, R. W., & Lees, B. G. (1993). Assessing the classification accuracy of multisource remote sensing data. *Remote Sensing of the Environment, 47*(3), 1–25.

Fitzgerald, R. W., & Lees, B. G. (1994). Spatial context and scale relationships in raster data for thematic mapping in natural systems. *Advances in GIS research, 1994*(1), 462–475.

Fitzgerald, R. W., & Lees, B. G. (1996). Temporal context in floristic classification. *Computers & Geosciences, 22*(9), 981–994.

Goodchild, M., Zhang, J., & Kyriakidis, P. (2009). Discriminant models of uncertainty in nominal fields. *Transactions in GIS, 13*(1), 7–23.

Hagen-Zanker, A., Straatman, B., & Uljee, I. (2005). Further developments of a fuzzy set map comparison approach. *International Journal of Geographical Information Science, 19*(7), 769–785.

Huang, Z., & Lees, B. (2005). Representing and reducing error in natural-resource classification using model combination. *International Journal of Geographical Information Science, 19*(5), 603–621.

Huang, Z., & Lees, B. G. (2004). Combining non-parametric models for multisource predictive forest mapping. *Photogrammetric Engineering and Remote Sensing, 70*(4), 415–425.

Huang, Z., & Lees, B. G. (2007). Assessing a single classification accuracy measure to deal with the imprecision of location and class: Fuzzy weighted Kappa versus Kappa. *Journal of Spatial Science, 52*(1), 1–13.

Larkin, J. H., & Simon, H. A. (1987). Why a diagram is (sometimes) worth ten thousand words. *Cognitive science, 11*(1), 65–100.

Lemon, O., & Pratt, I. (1999). Putting channels on the map: Verisimilitude and spatial constraints in a semantics of geographic information systems. *Logic, Language, and Computation, 2*, 143–164.

Lees, B. G. (1994). Decision trees, artificial neural networks and genetic algorithms for classification of remotely sensed and ancillary data. In *7th Australasian remote sensing conference, remote sensing and photogrammetry association Australia Ltd. Floreat* (pp. 51–60).

Lees, B. G. (1996a). Sampling strategies for machine learning using GIS. In M. F. Goodchild, L. Steyart, B. Parks, M. Crane, C. Johnston, D. Maidment & S. Glendinning (Eds.), *GIS and environmental modelling: Progress and research issues* (pp. 39–42). Fort Collins, Co: GIS World Inc.

Lees, B. G. (1996b). Improving the spatial extension of point data by changing the data model. In M. Goodchild (Ed.), *Proceedings of the Third international Conference on Integrating GIS and Environmental Modeling Santa Fe, New Mexico*, National Centre for Geographic Information and Analysis, WWW; CD.

Lees, B. G., & Ritman, K. (1991). Decision-tree and rule-induction approach to integration of remotely sensed and GIS data in mapping vegetation in disturbed or hilly environments. *Environmental Management, 15*(6), 823–831.

Mark, D. M., & Csillag, F. (1989). The nature of boundaries on 'area-class' maps. *Cartographica. Cartographica: The International Journal for Geographic Information and Geovisualization, 26*(1), 65–78.

McBratney, A. B., Santos, M. L. M., & Minasny, B. (2003). On digital soil mapping. *Geoderma, 117*(1), 3–52.

McHarg, I. L., & Mumford, L. (1969). *Design with nature* (p. 208), New York: Wiley.

Moore, D. M., Lees, B. G., & Davey, S. (1991). A new method for predicting vegetation distributions using decision tree analyses in a geographic information system. *Environmental Management, 15*(1), 59–71.

Openshaw, S. (1983). *The modifiable area unit problem*. Norwich, England: GeoBooks.

Peregrin, J. (2000). Fregean logic and Russellian logic. *Australasian Journal of Philosophy, 78*(4), 557–574.

Sinton, D. F. (1977). The inherent structure of information as a constraint to analysis: Mapped thematic data as a case study. International advanced study symposium on topological data structures for geographic information systems, Dedham, Mass. In G. Dutton (Ed.), *Harvard papers on geographic information systems, Harvard University* (p. 1978). Mass: Camb.

TWD, On-line discussion. (1997). *Thinking with diagrams*. http://www.mrc-cbu.cam.ac.uk/projects/twd/online-discussion. Accessed August 5, 2005.

van Niel, K., Laffan, S. W., & Lees, B. G. (2004). Error and uncertainty in environmental variables for ecological modelling: Sensitivity of variables to error in source elevation data. *Journal of Vegetation Science, 15,* 747–756.

Part II
Web and Crowd Sourcing Spatial Data Mining

A Framework for Event Information Extraction from Chinese News Online

Shuang Wang, Yecheng Yuan, Tao Pei and Yufen Chen

Introduction

The rapid development of the Internet has led to an explosion of available information and online services. As a result, a variety of structured heterogeneous web sites have appeared on the Internet, making it more difficult for users to find valuable information they are interested in. Information Retrieval (IR), via search engines is one of the most successful solutions to the problem. Typically, a search engine gives users a series of relevant web pages based on the key words. However, the drawback of existing IR is most evident when encountering huge search results, e.g., Google returns about 3,700,000 web page links for the keywords "debris flow". It is impossible to explore each record manually and locate the valuable information for debris flow hazards from the huge amount of data. Thus, we need to find a powerful and intelligent way to enhance IR. In this paper, we try to address the problem of discovering emergent events from Chinese news sites to automatically extract the relevant information, e.g., time, location, disaster situation, to recognize the information into structured forms that are more suitable for other information processing tasks, such as data mining, answering question and so on.

S. Wang · Y. Chen
Zhengzhou Institute of Surveying and Mapping,
Zhengzhou 450052, People's Republic of China
e-mail: wangsh@lreis.ac.cn

Y. Chen
e-mail: cyfbeijing@163.com

Y. Yuan (✉) · T. Pei
State Key Laboratory of Resources and Environment Information System,
Chinese Academy of Science, Beijing 100101, People's Republic of China
e-mail: yuanyc@lreis.ac.cn

T. Pei
e-mail: peit@lreis.ac.cn

C. Zhou et al. (eds.), *Spatial Data Handling in Big Data Era*,
Advances in Geographic Information Science,
DOI 10.1007/978-981-10-4424-3_5

In this paper, we construct a framework for event information extraction from various unstructured data, especially Chinese news. (Note that unstructured data does not imply that the data is structurally incoherent, but rather that the information therein is encoded in such a way that makes it difficult for computers to interpret it directly (Moens 2006), e.g., HTML, DOC). A prototype system Emergency Event Information Extraction System (EEIES) has been developed based on this framework. The system can collect web pages, extract information on emergent events from them and store the extracted data in database, with links back to the original documents.

Related Work

In recent decades, a number of information extraction (IE) systems have been developed to meet the need of automatically analyzing natural language text (Agichtein and Gravano 2000; AKT project 2005; Buitelaar et al. 2008; Donaldson et al. 2003; Muin et al. 2005; Mangassarian and Artail 2007; Chau and Xu 2007; Mykowiecka et al. 2009; SEKT project 2003; Lee and Lee 2007). For example, in the biomedical domain, the amount of literature is increasing rapidly, MEDLINE (Medical Literature Analysis and Retrieval System Online) comprises more than 17 million abstracts and the NLM (National Library of Medicine) continues to add 12,000 new abstracts each week (Muin et al. 2005). This situation has led to the development of such systems to handle scientific articles, e.g., Donaldson (2003) designed an information extraction system to locate protein-protein interaction data in the literature using machine learning methods.

Four basic types of elements can be extracted by these systems: entities, attributes, facts and events (Feldman and Sanger 2009). Entities are the basic proper names mentioned in the text, such as names of people, genes, and geographic locations. Attributes are the features of the extracted entities; Mykowiecka (2009) described a rule-based information extraction system to find certain information such as the age of the patient and drug dose, from mammography reports and hospital records of diabetic patients. Facts are the relations that exist between entities, such as social relationships between peoples Chau and Xu (2007). Agichtein and Gravano (2000) propose a bootstrapping method to find relations from text with minimal human intervention. Events can be defined as activities or occurrences of interest in which entities participate. Events extraction is the most complicated and challenging task in an IE system, which integrates former extraction tasks and requires further domain analysis. A few English IE systems or projects for events have been constructed, including SOBA (SmartWeb Ontology-based Annotation), an ontology-based information extraction system, which can extract football information from the SmartWeb and integrate it into a coherent knowledge base (Buitelaar et al. 2008).

Due to the complexity of the Chinese language and limitations of NLP (natural language processing) technology, few mature Chinese IE systems have hitherto been developed. Lee (2003) proposes an ontology-based fuzzy event extraction agent for Chinese e-news summarization. Strictly speaking, this is a summarization system rather than an IE system and cannot identify target information at the word or phrase level. To overcome the disadvantage, we propose an event model and develop a prototype IE system to extract event information at the word level. Meanwhile, we attempt to minimize the effort of constructing extraction patterns by maximizing the utility of machine learning algorithms.

The structure of this article is organized as follows. Section "Event Model" describes the event model. Section "System Framework" presents an overview of the framework for event information extraction and focuses on the key components. Section "Experiment" illustrates our prototype system with an experiment. Finally, Section "Conclusion and Future Work" gives a brief evaluation and discusses the future work.

Event Model

Definition of Event

The term "event" has different meaning in different fields. The traditional linguistic definition of an event consists of two aspectual parts: states and actions (Bethard and Martin 2006). States describe situations that are static or unchanging for their duration, while actions describe situations that involve some internal structure. In Topic Detection and Tracking (TDT is a DARPA-sponsored initiative to investigate the state of the art in finding and following new events in a stream of broadcast news stories) task, events are sets of documents that described "some unique thing that happens at some point in time" (Allan et al. 1998). In TimeML, a rich specification language for event and temporal expressions in natural language text, the event is a cover term for situations that occur at an exact time or last for a period of time (Pustejovsky et al. 2003). In the Automatic Content Extraction program (ACE), which is directed at the extraction of information from audio and image sources in addition to pure text, although this research effort is restricted to information extraction from text), an event is a complex structure with ancillary temporal information (Ahn 2006).

In this paper, we extend the term "event" defined in ACE: an event is an activity that occurs at a certain place and time, and in which one or more named entities (event roles) participated. So, place, time and entities are the features of an event. In general, these features can be divided into two types: the first includes the basic characteristic shared by all events, e.g., when and where events take place, while the second comprises the specific characteristics, which tag particular kinds of event, e.g., the magnitude of an earthquake. Therefore, we use a multi-layer object-oriented architecture to portray the event model, in which events can inherit from basic events or be composed of basic events. Details are described in the next section.

Architecture of Events

There are three layers in our multi-layer object-oriented architecture, including a conception layer, basic category layer and extended category layer. Figure 1 shows the structure of the event model.

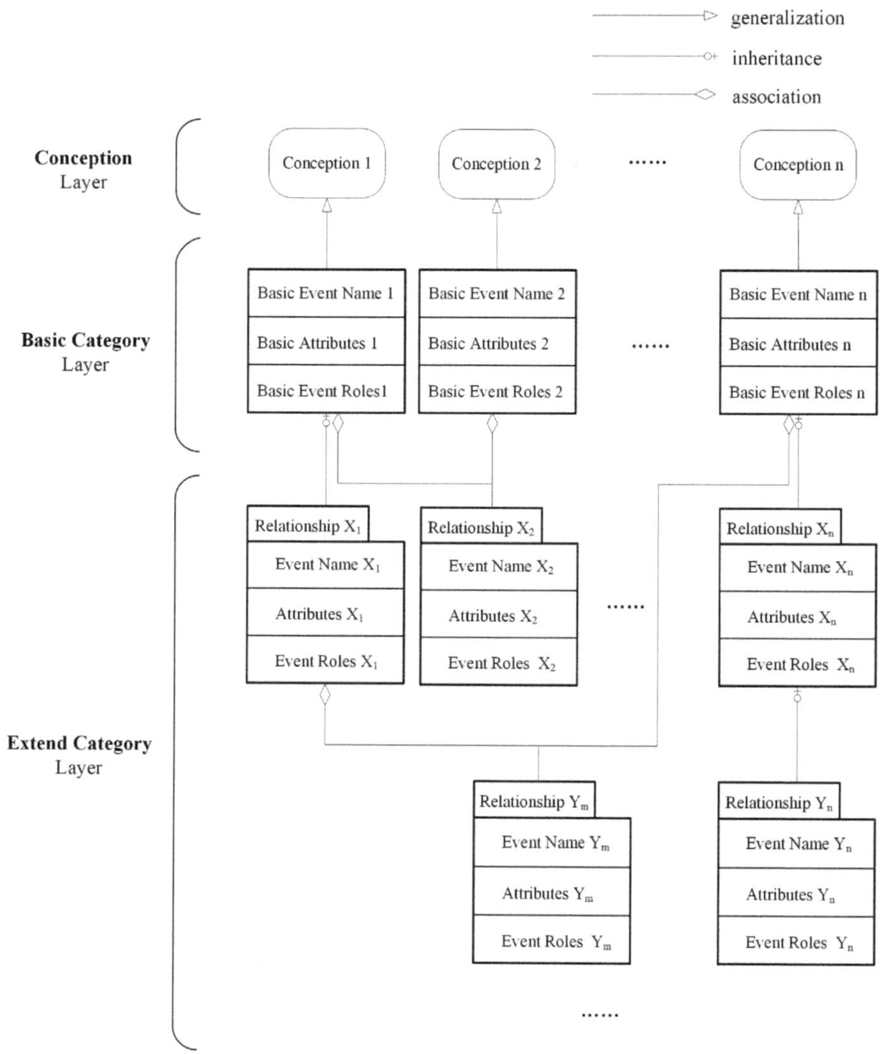

Fig. 1 Multi-layer object-oriented architecture of the event model

- Conception: the top layer of the architecture, which can be referred to as the event type. In this layer, each type of event is expressed by a set of widely acceptable terms in natural language.
- Basic Category: the second layer. Each basic category corresponds to a conception. Event name, basic event roles and attributes are contained in this layer. Thus, each conception is generalized by its own basic category.
- Extended Category: more event types can be formed by extending basic ones or combining them. Each extended category in this layer is a complex event that has more specific attributes and a relationship between itself and the basic categories. We define two relationships: inheritance and association. Inheritance represents the 'ancestor-posterity' relationship, while association denotes the 'combination' relationship.

Figure 2 shows an example of the architecture.

Fig. 2 Example of an object-oriented event model

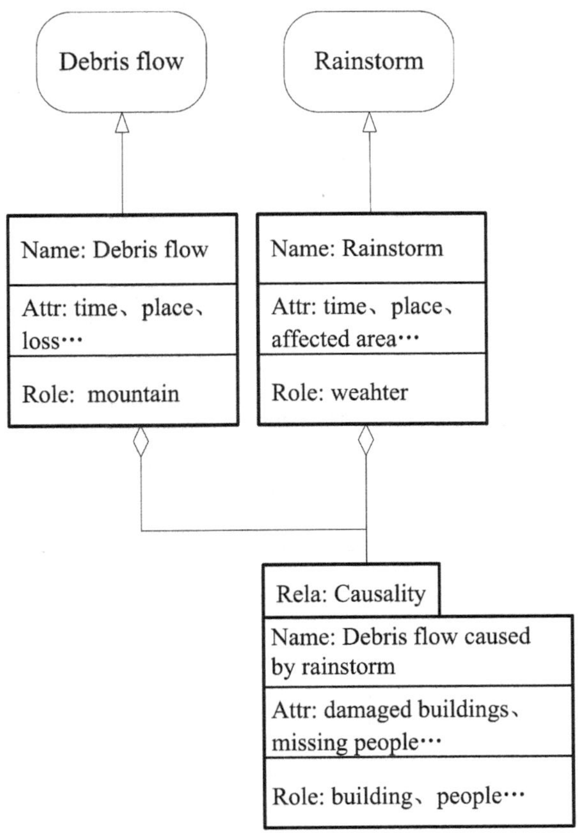

System Framework

Overview

The prototype system for extracting emergency event information consists of a Data Retrieval Agent (DRA), Document Processing Agent (DPA), Information Extraction Agent (IEA), and Knowledge Base (KB). The DRA is in charge of collecting data from various sources, including the Internet, Intranet, and local file systems. The DPA identifies potential emergency related contents from the data provided by the DRA. Then, the IEA determines important information from the contents with the help of the KB and stores it in a database for further study. Figure 3 shows the architecture of event extraction system.

Data Retrieval

Consisted of spider and an indexer, the data retrieval agent collects original data from various sources as system inputs. The goal of the spider is to visit many web sites and efficiently label as many as possible useful web pages. In our prototype system, to improve the validity of crawling, spiders are limited to a list of news web sites, such as Sina (http://news.sina.com.cn/), Sohu (http://news.sohu.com/), and

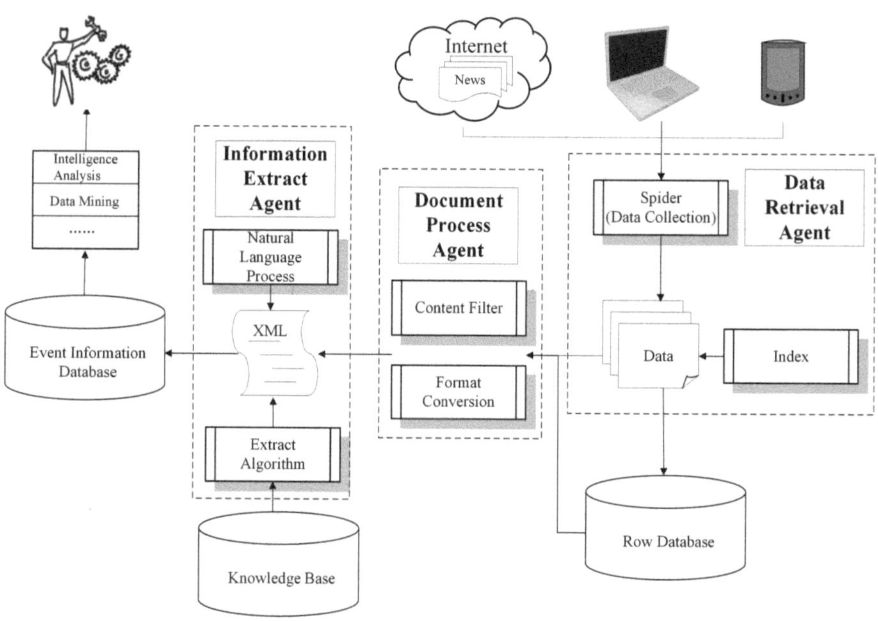

Fig. 3 Architecture of the event extraction system

Xinhua net (http://www.xinhuanet.com/). The indexer assigns each web page or document a unique doc-ID and creates indices for them. Finally, these documents are saved in database with their own URL. There have been many papers explaining how to construct a spider or indexer (Anh and Moffat 2005; Boldi et al. 2004; Cho et al. 1998; Carmel et al. 2001; Shkapenyuk and Suel 2002). Readers should refer to these references for further details.

Document Preprocessing

The goal of document preprocessing is to select potentially relevant web pages or other documents that may contain target event information and convert them into the uniform XML format. (See Fig. 4). Since most documents or pages are irrelevant to the event being extracted, it is necessary to remove them before they are sent to the information extraction agent. A content filter has been designed to carry out this task automatically, which up the domain terms in glossaries stored in the knowledge base in advance; only those containing the domain terms are kept as

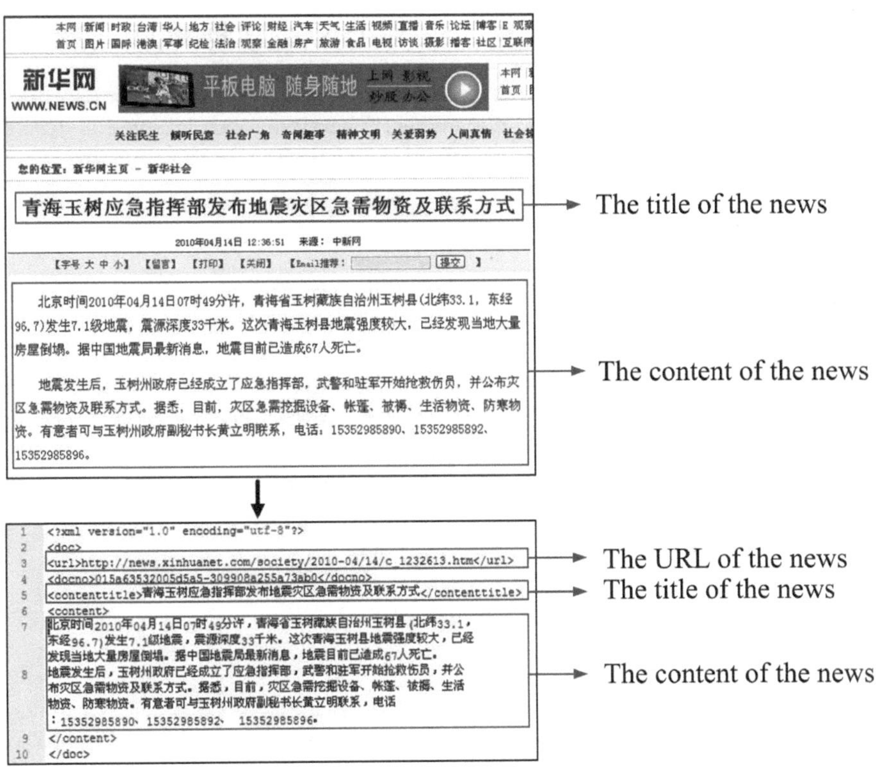

Fig. 4 Document preprocessing

candidate documents. To maintain completeness of the potentially relevant documents, we use no more than three domain terms. Taking "earthquake" as an example, the content filter uses the term "earthquake" to select documents that may contain information about an earthquake.

Another important function of the content filter is to find the main text body within the web page. A typical news web page contains many elements besides the actual news text body, such as navigation information, links to other sites, and advertisement. The wrapper induction method (Crescenzi et al. 2001; Freitag and Kushmerick 2000; Kushmerick 2000) which generates wrappers by the generalization of a set of examples of (page, content) pairs is introduced to tag user-interested content embedded in HTML pages. Essentially, a wrapper class is a mapping function, which maps the kind of pages to the content hidden in such pages. Considering that the DRA only visits the given web sites, we develop a series of wrappers to analyze the web structure of these sites. Our experiment shows that over 90% of pages can be handled correctly. Obviously, the scalability and reusability of these wrappers are poor, something which we aim to improve in our future work.

Finally, these 'clean' web pages or documents are transformed into XML format. The reason for using XML is as below. HTML is designed to "exhibit" data, and not to "process" data, so the tags used in HTML are limited and lack extensibility. Besides, the structure of HTML is not strict, making it difficult to validate the structure, but XML eliminates these constrains. XML supports user-defined tags and requires a comparatively strict logical structure, which facilitates the execution of programs to parse the XML documents. Therefore, we designed a format transformation algorithm to convert the files into XML. The main elements in XML include the title, doc-ID, URL, and text content.

Information Extraction

The information extraction agent plays a major role of in the whole system. Information extraction is a complex task that requires collaboration with natural language processing, machine learning, pattern recognition, etc. Therefore, a series of dependent steps, which can be divided into two major components, namely, natural language processing and pattern analysis, are applied sequentially. Figure 5 depicts a flow chart of the information extraction agent.

Natural Language Processing

The goal of natural language processing is to ascertain the lexical and syntax features of each sentence in a document. These features will later be used in pattern analysis. Since each language has its own unique characters, natural language processing differs for each. For example, it is fairly easy to identify word

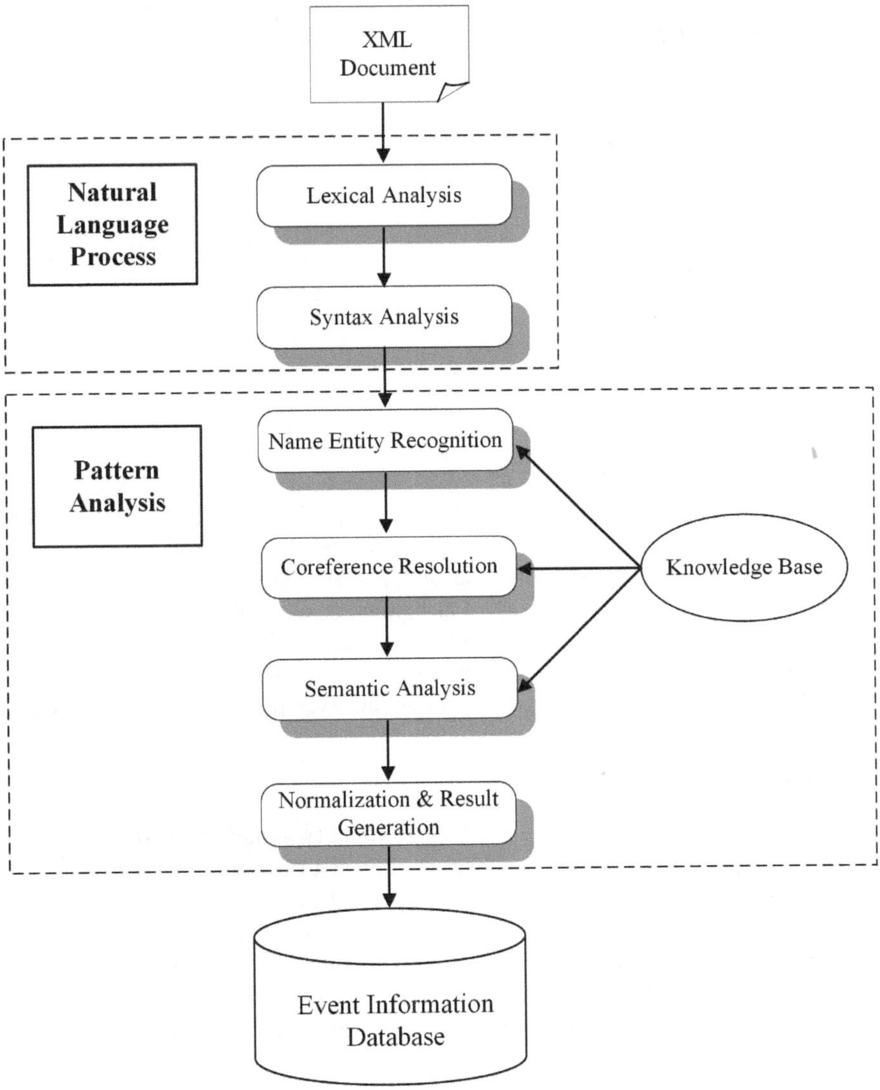

Fig. 5 Flow chart of information extraction agent

boundaries in European languages, but much harder in Chinese, because there is no Chinese character representing the space for separating words. Thus, the process begins with lexical analysis.

- ***Lexical Analysis***

There are many Chinese word segmentation models, which can be divided into two categories: handcrafted and statistics-based. Previous studies show that the latter

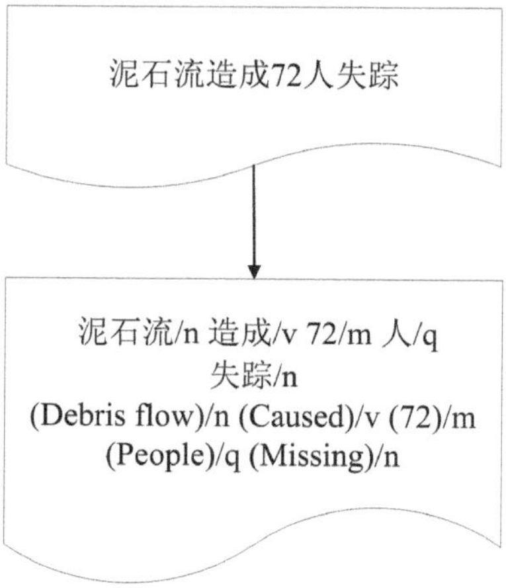

Fig. 6 Example of lexical analysis result by ICTCLAS

category yields better performance than the former (HUANG and ZHAO 2007). For our system, we selected the ICTCLAS (Institute of Computing Technology, Chinese Lexical Analysis System), one of the best segmentation systems, as the Chinese lexical analysis tool (Zhang et al. 2003a, b, c). ICTCLAS uses a HHMM (hierarchical hidden Markov model) and incorporates Chinese word segmentation, Part-Of-Speech tagging, disambiguation and unknown words recognition into an entire theoretical framework. Figure 6 shows an example of the lexical analysis results using ICTCLAS.

- *Syntax Analysis*

The syntax parser analyzes entire sentence to identify syntactic relations between the heads of phrases. We used the Stanford Parser to determine the grammatical structure of sentences. The Stanford Parser is a lexicalized probabilistic parser developed by the Stanford natural language processing group; it implements a factored product model with a separate PCFG (probabilistic context-free grammars) phrase structure and lexical dependency expert (Klein and Manning 2003a, b). Figure 7 shows an example of a phrase structured grammar created by the Stanford Parser.

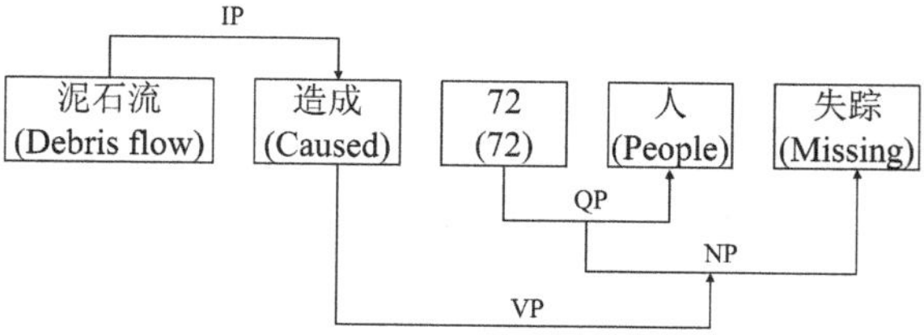

Fig. 7 Phrase structure grammar of the sentence "泥石流 (debris flow) 造成 (caused) 72(72) 人 (people) 失踪 (missing)"

Pattern Analysis

A robust, reliable, self-learning extraction pattern is the key to the information extraction system. Previous systems made use of handcrafted algorithms (Appelt et al. 1993; Lehnert et al. 1991; Soderland et al. 1995), which were crafted by linguists in cooperation with domain experts. Handcrafted methods usually take a long time to perfect the rule set resulting in poor adaptability. Modern systems employ machine-learning techniques, which automatically induce rules by studying a set of annotated training examples (Ciravegna 2001). In the last few years, several statistics-based models (supervised learning models) have been developed (Isozaki and Kazawa 2002; Peng and McCallum 2006). However, since it is so costly to obtain an annotated corpus researchers have recently become more interested in exploring semi-supervised (or "weakly supervised") techniques (Riloff and Jones 1999; Yarowsky 1995).

In our prototype system, due to the lack of human labor to create a large-scale annotated corpus, we adopted a rule-based method for each subtask in Fig. 5 instead of a statistics-based method. The procedure is explained as follows.

(1) **Assigning Time-stamps and Place to Events (Named Entity Recognition)**

Time and place are important clues for tracking the evolution of an event. Identifying the location and temporal expressions that has been involved in Chinese lexical analysis process belongs to the traditional named entity recognition task. After word segmentation by ICTCLAS, we can correctly obtain most temporal words, but no more than two thirds of the location words on average. To improve the location words recognition rate, we added a gazetteer to amend the original segmentation errors. Here, we pay special attention to how to map each temporal word to a calendar and link each location word to a map.

- ***Time***

There are two kinds of temporal words: absolute and relative. "Absolute" means that the temporal word itself can be mapped to a calendar without other supplementary information, e.g. "2010年8月2日" (08/02/2010). "Relative" means the word needs a reference time to determine its place in the calendar, such as "2小时前" (two hours ago), "昨天" (yesterday). Thus, it is not an easy job to calculate the true time from the relative temporal expressions; the reference time and the relationship between them needed to be find out. Allen (1991) summarizes 13 kinds of temporal relationships. In this paper, we simplify them to 4 relationships: equal to, contains, before and after Listed in Table 1.

Table 1 Four temporal relationships

Relationship	Symbol	Legend
X before Y	<	XXX YYY
X equal to Y	=	XXX YYY
X contains Y	C	XXXXX YYY
X after Y	>	YYYXXX

X temporal word; *Y* reference time

General speaking, the first sentence of each report contains time information. For example:

"中新网5月30日电　水利部部长……"

(30/5 Chinanews report Minister of Water Resources…)

We take this date which is in bold as the report time of the particular news item. When analyzing relative temporal expressions in articles, we take the absolute temporal expression appearing in the same sentence or the last one as the reference time. If both of these are missing, then the report time of the article is substituted. Once the reference time and temporal relationship have been determined, we can calculate their calendar date according to the reasoning rules.

The time format used in our prototype system is as follows:

$$\mathbf{T-stamp}:\{\mathbf{YYYY-MM-DD-HH-NN-SS}:\ \mathbf{W}:\ \mathbf{A}\}$$

YYYY-year, **MM**-month, **DD**-day, **HH**-hour, **NN**-minute, **SS**-second, **W**-number of the day in the week (1- Monday … 7- Sunday), and **A**-special day's name, e.g., "元旦 (New Year's day). In fact, only a few temporal expressions can be precisely assigned to the unit of second (typically these can be assigned to a day), and therefore, we may not find all the values for each argument of the time format, but at least some of them can be identified.

- *Place*

It is not sufficient to identify the entity names, what's more important is to match these names to real places. To achieve this target, we defined a structure containing eight sub-types for locations in our prototype system:, namely, **country**, **province**, **city**, **county**, **road**, **river**, **mountain**, and **island**. We aim to identify all names and place them in the correct positions in the structure. To improve precision of location recognition, we utilized a national gazetteer with a scale of 1: 1,000,000 ("全国 1:100 万地名库") provided by the National Fundamental Geographic Information System. Most commonly used geographic entity names can be found in the gazetteer.

The ambiguity of names is the main cause of errors. There are two types of ambiguity: different places sharing the same name and one name having different semantic meanings in different contexts. For example, "北京 (Beijing)" means the city of Beijing in "2008年奥运会在北京举行" (2008 Olympic Games were held in Beijing), but implies the Chinese Government in "北京表示反对抵制奥运行为" (Beijing is against the behavior of resisting the Olympic Games). To solve the ambiguity, we need context information.

(2) **Co-reference Resolution**

Co-reference resolution is the process of determining whether two expressions in different sentences refer to the same entity, especially the same event in our study. For example, in the following sentence, the expressions are depicted in bold.

"The **rainstorm** itself wasn't that exceptional in terms of appearance on radar and satellite, but it did cover a fairly large area."

Here, "**it**" refers to the "rainstorm", not "radar" or "satellite".

In our current co-reference model, we have built a glossary for event reference, such as "灾难" (disaster), "事故" (calamity). These non-event noun phrases (not currently in the glossary) have not yet been implemented. Given a noun phrase, the model searches for the most recently mentioned entities with have the same event class or a super-class or a sub-class with the noun phrase. If such an entity exists, it is taken as the reference for the noun phrase; otherwise, a new entity is defined.

(3) **Event Attributes Recognition (Semantic Analysis)**

This step links the identified temporal expressions, phrases of location, and other attribute information to their own events. We consider event attribute recognition to be a semantic relation analysis problem.

- *Semantic Relations*

Semantic relations reveal the domain-specific relationship between pairs of phrases or words, e.g., in the sentence "泥石流造成72人失踪" (There are 72 people missing in this debris flow), the semantic relationship between "泥石流" (debris flow) and "失踪" (missing) is referred to as "Cause-Missing".

A semantic relation consists of the relation name, head phrase, modifier phrase, modality, and polarity. Table 2 gives an example.

Modality and polarity are taken from ACE's definition for event attributes. Modality has two values: asserted and other. "Asserted" means that the relationship in the sentence is in the affirmative tone, e.g., "泥石流卷走了30人" (The debris flow covered 30 people). "Other" refers to "Believed Events/Hypothetical Events/Commanded and Requested Events" and so on. Polarity also has two values: negative and positive. "Negative" means the action did not happen, e.g. "泥石流没有造成人员伤亡" (There were no people injured in this debris flow).

Semantic relations correspond to the attributes of the event model. They are joined as a result of sharing the same phrases, e.g., the phrase "失踪" (missing) in Table 2 is the conjunction that connects the relation "Cause-Missing" and "Missing-Number". All attributes of the event can be described by semantic relations.

- *Semantic Patterns*

For our system, we proposed a machine-learning algorithm that learns information extraction patterns from hand-tagged examples. Machine learning is a positive

Table 2 Example of the semantic relation for the sentence "泥石流造成72人失踪" (72 people missing due to debris flow)

Relation name	Head phrase	Modifier phrase	Modality	Polarity
Cause-Missing	泥石流(debris flow)	失踪(missing)	asserted	positive
Missing-Number	失踪(missing)	72人(72 people)	asserted	positive

Table 3 Semantic pattern for "Cause-Missing"

Index	Word	Word type	POS	Lexical dependency
1	泥石流(debris flow)	Event name	N	IP(1,3)
2	*	#	#	#
3	造成(cause)	Trigger	V	VP(3,5), IP(1,3)
4	*	#	#	#
5	失踪(missing)	Missing	N	VP(3,5)

example-covering algorithm that tries to cover as many positive examples as possible, while not covering any negative examples. We used a linked chain model to represent the extraction patterns for semantic relations. The model is defined as a list consisting of a series of word nodes; each node carries a few attributes, including word content, word type, POS, and a lexical dependency relation between them. Table 3 gives an example.

The character "*" in the pattern can match any number of words, i.e. any reasonable words can appear between "泥石流" (debris flow) and "造成" (cause). This pattern will match a sentence like "突如其来的泥石流致使多人下落不明" (The sudden debris flow resulted in many people being missing). Here, node 1 matches a noun phrase with the semantic type "event name", that is, a noun phrase whose head is a word containing the "event name", e.g., "泥石流" (debris flow). The semantic classes of words are pre-defined in a domain-specific semantic glossary in the knowledge base. Similarly, node 2 matches a verb phrase whose head is a "trigger" class, e.g., "致使" (result in), node 3 matches a noun phrase whose head is the "missing" class, e.g., "下落不明" (disappear). Besides, the lexical dependency between nodes 2 and 3 must be a VP with node 1 the subject.

The use of the word semantic class broadens the coverage of the patterns. Meanwhile, the coverage is restricted by the lexical dependencies between these words or phrases. This method can ignore the modified words or phrases around the nodes, e.g., "突如其来" (sudden), "多人" (many people) in the example, thereby guaranteeing the consistency of the sentence structure at the same time. The patterns are learned from an annotated training corpus using the induction algorithm, while event information can be extracted by applying these patterns to new articles.

(4) **Normalization and Result Generation**

We chose a template in XML format as the information extraction output. The template was designed in advance based on the event model described in Section "Event Model". The elements in the template are the attributes of the event, which should be expressive enough to capture important details. Figure 8 gives an example of the template for an earthquake. The elements in this template include time, place, magnitude, number of injured people, number of death, and so on.

Extracted data should be normalized before being filled into the placeholders of the template, because writing styles differ from document to document, e.g.,

```xml
<?xml version="1.0" encoding="utf-8"?>
<doc>
<url>http://news.xinhuanet.com/society/2010-04/14/c_1232613.htm</url>
<docno>015a63532005d5a5-309908a255a73ab0</docno>
<contenttitle>青海玉树应急指挥部发布地震灾区急需物资及联系方式</contenttitle>
<content>
北京时间2010年04月14日07时49分许，青海省玉树藏族自治州玉树县(北纬33.1，
东经96.7)发生7.1级地震，震源深度33千米。这次青海玉树县地震强度较大，已经
发现当地大量房屋倒塌。据中国地震局最新消息，地震目前已造成67人死亡。
地震发生后，玉树州政府已经成立了应急指挥部，武警和驻军开始抢救伤员，并公
布灾区急需物资及联系方式。据悉，目前，灾区急需挖掘设备、帐篷、被褥、生活
物资、防寒物资。有意者可与玉树州政府副秘书长黄立明联系，电话
：15352985890、15352985892、 15352985896。
</content>
</doc>
```

```xml
<?xml version="1.0" encoding="utf-8"?>
<event type="地震" docno="015a63532005d5a5-309908a255a73ab0">
<time yyyy="2010" mm="4" dd="14" hh="7" nn="49" ss="" w="" a="" p="at"
modality="asserted" polarity="positive"/>
<place name="玉树" type="county" modality="asserted" polarity="positive"/>
<dzjs value="7.1" unit="级" modality="asserted" polarity="positive"/>
<death value="67" unit="人" modality="asserted" polarity="positive"/>
</event>
```

Fig. 8 Example of the event template for an earthquake

numbers in news articles may be in DBC case ("123") or in SBC case ("123"), even in Chinese format ("一二三" or "壹贰叁"). A similar problem exists in time and event names. Normalization for the time has been discussed in Section "Pattern Analysis", where each temporal phrase is mapped to a calendar date. An event name can be indicated by a number of names or aliases, e.g., "洪灾", "洪涝灾害", "洪水" (flood). A series of mapping rules are used to normalize numbers and names in this system.

Knowledge Base

As is commonly known, almost every extraction process requires some "knowledge", e.g., a linguistic dictionary for lexical and syntax analysis, gazetteer and glossary for domain term for name entity recognition, and trigger patterns for information identification. All of these constitute the "knowledge base" in our system; the function of each has been discussed separately in a previous section. Setting up the KB is an expensive and time-consuming process, which has greatly restricted the application of information extraction.

Experiment

To validate the system, we design an experiment for extracting emergency events. 116,577 pieces of news were collected from the given web sites in May and June, 2008. In the document preprocessing, the term "洪灾" (flood) is used as key word to select articles on flood, and 366 articles were selected in all. Nevertheless, some of them were duplicates and some were irrelevant. For example, an article with the title of "朱广沪热评欧锦赛:雇佣军不是洪水猛兽" is actually a football report; here "洪水猛兽" (flood) is an analogy for fearful things, not a real flood. Table 4 lists the details.

We designed six semantic relations for flood: "Happen-Time" (when the flood happened), "Happen-Place" (where the flood happened), "Cause-Missing" (are people missing in the flood), "Cause-Death" (have people died in the flood), "Missing-Number" (how many people are missing in the flood), and "Death-Number" (how many people died in the flood). We manually annotated the relations for each article, including semantic relation type, head phrase, modifier phrase, modality, polarity and so on. Figure 9 illustrates a fragment of an annotated article.

We used the 39 articles from May as the training data, and the 167 articles from June as the test data. Our system learned extraction patterns from instances annotated in the training data. Then, the patterns were applied to the test data. We evaluated the efficiency of the system by comparing actual recognized relation instances to the correct result. Figure 10 depicts the flow chart of the experiment. Table 5 gives the details the results of the experiment.

Precision and recall are defined as in Sebastiani (2002):

Table 4 Articles selected for the term "洪灾" (flood)

Event type	May, 2008 (training data)			June, 2008 (test data)		
	Relevant	Irrelevant	Duplicated	Relevant	Irrelevant	Duplicated
Flood	39	9	11	167	79	61

	A	B	C	D	E	F	G	H	I	J
1	Content	Word POS	Semantic Relation	Num	Is head phrase	Modified phrase index	Is modifier phrase	Head phrase index	Modality	Polarity
20	截止	vi								
21	到	p								
22	２７日	t	Happen-Time	1			Yes	26	asserted	positive
23	１１时	t	Happen-Time	1			Yes	26	asserted	positive
24	，	wd								
25	此次	rz								
26	洪灾	n								
27	已	d								
28	造成	v								
29	9	m	Death-Numer	2	Yes	31			asserted	positive
30	人	n								
31	死亡	vi	Cause-Death	3			Yes	26	asserted	positive
32	１１	m	Missing-Number	4	Yes	34			asserted	positive
33	人	n								
34	失踪	vi	Cause-Missing	5			Yes	26	asserted	positive
35	。	wj								

Fig. 9 Fragment of an annotated article

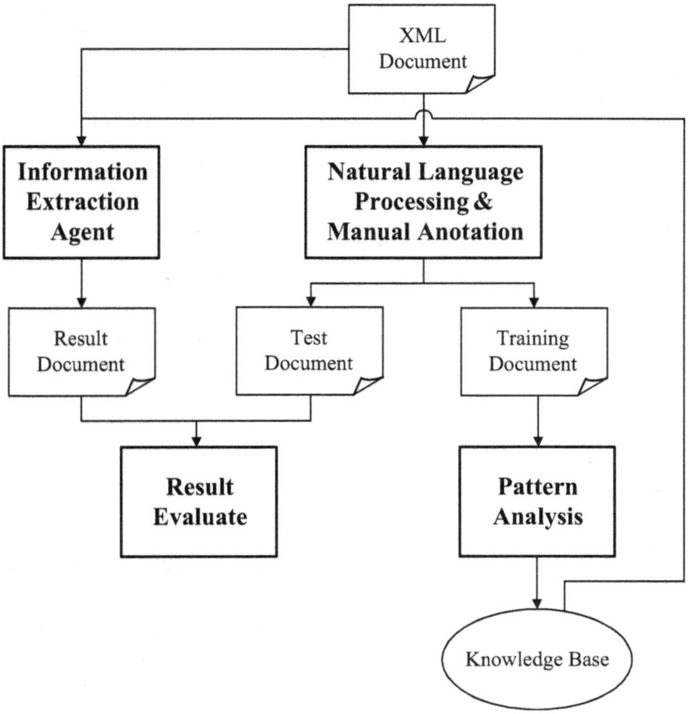

Fig. 10 Flow chart of the experiment

Table 5 Experimental results in terms of precision, recall, and F-measure

Relation types	Actual instances	Precision (%)	Recall (%)	F-measure (%)
Happen-time	271	85.9	74.6	80.25
Happen-place	167	82.1	77.1	79.6
Cause-missing	153	87.6	82.3	84.95
Cause-death	144	87.4	84.8	86.1
Missing-number	147	88.3	85.6	86.95
Death-number	132	89.2	87.4	88.3

$$Precision = \frac{COR}{COR + INC} \quad (1)$$

$$Recall = \frac{COR}{COR + MIS} \quad (2)$$

where COR denotes the correctly recognized relations, INC the incorrect ones, and MIS the missing ones.

The F-measure is the harmonic mean between precision and recall:

$$F = \frac{(\beta^2 + 1.0) \times P \times R}{(\beta^2 \times P) + R} \quad (3)$$

where β is the argument of weight, in our experiment, $\beta = 1$.

The result shows that precision for all types is relatively higher than recall. This means our system's ability to recognize is better than its covering. Although the extraction task in our experiment is not complex, we think that this result is acceptable and that the framework is feasible enough for application in practice.

Conclusion and Future Work

In this paper, we presented a methodology for automatically extracting event information from Chinese news online and storing the extracted results in machine-readable XML for further application. A prototype system, EEIES, comprising several sub-components namely, DRA, DPA, IEA, and KB, was developed to perform event information extraction. The Chinese lexical analyzer ICTCLAS, and syntax analyze Stanford Parser were embedded in our system for natural language processing. Furthermore, we proposed an object-oriented event model to capture the important information.

We presented a method that can automatically discover patterns by learning from hand-annotated training examples with the help of a knowledge base. The major contribution of the method is the ability to classify the semantic classes of words or phrases and identify the linguistic patterns that express semantic relationships.

We conducted an experiment in extracting information (time, place, population of missing etc.) of flood disasters from Chinese news online. The experiment showed a high recall and precision. The results indicate that this framework is feasible in practice, and also demonstrate that our system is a powerful supplement to IR. Besides, the extracted information with semantic tag in XML can be used in other applications, e.g., answering questions, data mining, and intelligence analysis.

Future work will be focused on the following aspects. (1) Exploring more statistics-based methods and integrating other online knowledge bases into our system, such as HowNet (a large lexical database of Chinese) and Wikipedia, to support discovering hidden facts. (2) Extracting information from other data formats besides free text, e.g., tables, images, and videos. Although direct content extracting from images or videos is still a challenge problem, the captions (if available) provide a valuable resource for determining the content and can thus be utilized.

Acknowledgements The study was supported by the National Science Found for Distinguished Young Scholars of China (Grant No. 41525004) and the Science Fund for Creative Research Groups (Grant No. 41421001).

References

Agichtein, E., & Gravano, L. (2000). 'Snowball: Extracting relations from large plain-text collections'. *Paper Presented at Proceedings of the Fifth ACM International Conference on Digital Libraries*. New York, USA, June 02–07, 2000.

Ahn, D. (2006). 'The stages of event extraction'. *Paper Presented at Proceedings of the COLING/ACL 2006 Workshop on Annotating and Reasoning about Time and Events*. Sydney, Australia, July 23, 2006.

AKT project. (2005). *Advanced Knowledge Technologies project. Obtained through the Internet.* http://www.aktors.org/akt/. Accessed Feb 8, 2010.

Allan, J., Carbonell, J., Doddington, G., Yamron, J., & Yang, Y. (1998). 'Topic detection and tracking pilot study: Final report', *Paper Presented at Proceedings of the DARPA Broadcast News Transcription and Understanding Workshop*. April 1998.

Allen, J. (1991). Time and time again: The many ways to represent time. *International Journal of Intelligent Systems, 6*(4), 341–355.

Anh, V. N., & Moffat, A. (2005). Inverted index compression using word-aligned binary codes. *Information Retrieval, 8*(1), 151–166.

Appelt, D., Hobbs, JR., Bear, J., Israel, D., & Tyson, M. (1993). 'FASTUS: A finite-state processor for information extraction from real-world text'. *Paper Presented at the Proceedings of the 13th International Joint Conference on Artificial Intelligence IJCAI-93. August 1993.* Chambery, France.

Bethard, S., & Martin, J. H. (2006). 'Identification of event mentions and their semantic class'. *Paper Presented at Proceedings of the 2006 Conference on Empirical Methods in Natural Language Processing EMNLP 2006*. Sydney, Australia, July 22–23, 2006.

Boldi, P., Codenotti, B., Santini, M., & Vigna, S. (2004). 'Ubicrawler: A scalable fully distributed web crawler'. *Software: Practice and Experience, 34*(8), 711–726.

Buitelaar, P., Cimiano, P., Frank, A., Hartung, M., & Racioppa, S. (2008). Ontology-based information extraction and integration from heterogeneous data sources. *International Journal of Human-Computer Studies, 66*(11), 759–788.

Carmel, D., Cohen, D., Fagin, R., Farchi, E., Herscovici, M., Maarek, Y. S., et al. (2001). 'Static index pruning for information retrieval systems'. *Paper Presented at Proceedings of the 24th Annual SIGIR Conference on Research and Development in Information Retrieval. New Orleans, USA*, September 9–13, 2001.

Chau, M., & Xu, J. (2007). Mining communities and their relationships in blogs: A study of online hate groups. *International Journal of Human-Computer Studies, 65*(1), 57–70.

Cho, J. H., Garcia-Molina, H., & Page, L. (1998). 'Efficient crawling through URL ordering'. *Computer Networks and ISDN Systems, 30*(1–7), 161–172.

Ciravegna, F. (2001). 'Adaptive information extraction from text by rule induction and generalization'. In *The Proceedings of the 17th International Joint Conference on Artificial Intelligence IJCAI-01. Washington, USA*. August 4–10, 2001.

Crescenzi, V., Mecca, G., & Merialdo, P. (2001). 'Roadrunner: Towards automatic data extraction from large web sites'. *Paper Presented at Proceedings of the 27th International Conference on Very Large Data Bases. Rome, Italy*, September 11–14, 2001.

Donaldson, I., Martin, J. De, Bruijn, B., Wolting, C., Lay, V., Tuekam, B., et al. (2003). PreBIND and textomy—mining the biomedical literature for protein-protein interactions using a support vector machine. *BMC Bioinformatics, 4*(1), 11–24.

Feldman, R., & Sanger, J. (2009). *The text mining handbook advanced approaches in analyzing unstructured data*. Beijing: Posts & telecom press.
Freitag, D., & Kushmerick, N. (2000). 'Boosted wrapper induction'. *Paper Presented at Proceedings of the Seventeenth National Conference on Artificial Intelligence and Twelfth Conference on Innovative Applications of Artificial Intelligence. Austin, USA*, July 30–August 03, 2000.
Grishman, R., Huttunen, S., & Yangarber, R. (2002). Information extraction for enhanced access to disease outbreak reports. *Journal of Biomedical Informatics, 35*(4), 236–246.
Huang, C.-N., & Zhao, H. (2007). Chinese word segmentation: A decade review. *Journal of Chinese Information Processing, 21*(3), 8–19.
Isozaki, H., & Kazawa, H. (2002). 'Efficient support vector classifiers for named entity recognition'. In *Proceedings of the 19th International Conference on Computational Linguistics. Taipei, Taiwan*. August 24–September 01, 2002.
Klein, D., & Manning, C. D. (2003a). 'Fast exact inference with a factored model for natural language parsing'. In *Advances in neural information processing systems 15 NIPS 2002. Whistler, British Columbia, Canada*, December 10–12, 2002.
Klein, D., & Manning, C. D. (2003b). 'Accurate Unlexicalized Parsing'. In *Proceedings of the 41st Annual Meeting on Association for Computational Linguistics. Sapporo, Japan*, July 07–12, 2003.
Kushmerick, N. (2000). Wrapper induction: Efficiency and expressiveness. *Artificial Intelligence, 118*(1–2), 15–68.
Lee, C. S., Chen, Y. J., & Jian, Z. W. (2003). Ontology-based fuzzy event extraction agent for Chinese e-news summarization. *Expert Systems with Applications, 25*(3), 431–447.
Lee, S., & Lee, G.-G. (2007). Exploring phrasal context and error correction heuristics in bootstrapping for geographic named entity annotation. *Information Systems, 32*(4), 575–592.
Lehnert, W., Cardie, C., Fisher, D., Riloff, E., & Williams, R. (1991). 'University of Massachusetts: Description of the CIRCUS system as used for MUC-3'. In *Proceedings of Third Message Understanding Conference. San Diego, USA*, May 1991.
Mangassarian, H., & Artail, H. (2007). A general framework for subjective information extraction from unstructured English text. *Data & Knowledge Engineering, 62*(2), 352–367.
Moens, M.-F. (2006). *Information extraction: Algorithms and prospects in a retrieval context*. Dordrecht: Springer.
Muin, M., Fontelo, P., Liu, F., & Ackerman, M. (2005). SLIM: An alternative web interface for MEDLINE/PubMed searches—a preliminary study. *BMC Medical Informatics and Decision Making, 5*(1), 37.
Mykowiecka, A., Marciniak, M., & Kupsc, A. (2009). Rule-based information extraction from patients' clinical data. *Journal of Biomedical Informatics, 42*(5), 923–936.
Peng, F., & McCallum, A. (2006). Information extraction from research papers using conditional random fields. *Information Processing and Management, 42*(4), 963–979.
Pustejovsky, J., Castano, J., Ingria, R., Saurí, R., Gaizauskas, R., Setzer, A., & Katz, G. (2003). 'TimeML: Robust specification of event and temporal expressions in text'. *Paper Presented at Proceedings of the IWCS-5 Fifth International Workshop on Computational Semantics*. Tilburg, Netherlands, January 15–17, 2003.
Riloff, E., & Jones, R. (1999). 'Learning dictionaries for information extraction by multi-level bootstrapping'. In *Proceedings of the Sixteenth National Conference on Artificial Intelligence and the Eleventh Innovative Applications of Artificial Intelligence Conference Innovative Applications of Artificial Intelligence. Orlando, Florida, United States*. July 18–22, 1999.
Sebastiani, F. (2002). Machine learning in automated text categorization. *ACM Computing Surveys CSUR, 34*(1), 1–47.
SEKT project. (2003). Semantic Knowledge Technologies project. Obtained through the Internet: http://www.sekt-project.com/. Accessed Feb 8, 2016.
Shkapenyuk, V., & Suel, T. (2002). 'Design and implementation of a high-performance distributed web crawler'. *Paper Presented at Proceedings of the 18th International Conference on Data Engineering. San Jose, USA*, February 26–March 01, 2002.

Soderland, S., Fisher, D., Aseltine, J., & Lehnert, W. (1995). 'CRYSTAL: Inducing a conceptual dictionary'. In *Proceedings of the 14th International Joint Conference on Artificial Intelligence IJCAI-95. Montreal, Canada*, August 20–25, 1995.

Yarowsky, D. (1995). 'Unsupervised word sense disambiguation rivaling supervised methods'. In *Proceedings of the 33rd Annual Meeting on Association for Computational Linguistics. Cambridge, Massachusetts*. June 26–30, 1995.

Zhang, H.-P., Liu Q., Cheng, X.-Q., Zhang, H., & Yu, H.-K. (2003a). 'Chinese lexical analysis using hierarchical hidden Markov model'. *Paper Presented at Proceedings of the Second SIGHAN Workshop on Chinese Language Processing. Sapporo, Japan*. July 11–12, 2003.

Zhang, H., Liu, Q., Yu, H., Cheng, X., & Bai, S. (2003b). Chinese named entity recognition using role model. *Computational Linguistics and Chinese Language Processing, 8*(2), 29–60.

Zhang, H.-P., Yu, H.K., Xiong, D.-Y., & Liu, Q. (2003c). 'HHMM-based Chinese lexical analyzer ICTCLAS'. *Paper Presented at Proceedings of the Second SIGHAN Workshop on Chinese Language Processing. Sapporo, Japan*, July 11–12, 2003.

Evaluating Neighborhood Environment and Utilitarian Walking Behavior with Big Data: A Case Study in Tokyo Metropolitan Area

Hao Hou and Yuji Murayama

Introduction

Fast development in transport technology has brought great convenience to people's daily life, especially for those who live in highly urbanized areas. However, the convenience in daily life caused that a significant proportion of people all over the world adopted a physically inactive lifestyle (Van Dyck et al. 2013). Around 50% of the population in America were found to be physically inactive (Hallal et al. 2012) and the proportion of inactive adults in Australia even reached 57% (Wang et al. 2016). Besides the high proportion, the trend of an increase in the proportion of physically inactive people was also noticed. In Japan, the proportion of adults achieving 10,000 steps per day fell by 5% from 2000 to 2007 (Inoue et al. 2011). The evidence in China showed that the average physical activity level of Chinese adults decreased by more than 30% from 1999 to 2006 (Ng et al. 2009). Physical inactivity was found to be linked with higher risks of overweight and obesity. Besides, physical inactive lifestyles affect people's mental health as it can increase the mental pressure and cause depression (Wang et al. 2016). As a result, the promotion of physical activity is attracting high attention and becoming a health priority in recent years (Heath et al. 2012).

Among all the physical activities, walking is recognized as one of the most common, accessible, inexpensive forms of physical activity and is an important component of total physical activity in adult populations (Hallal et al. 2012). In this context, knowledge on how to promote people's daily walking behavior is critical. In recent years, a number of studies have revealed the relationships between

H. Hou (✉) · Y. Murayama
Division of Spatial Information Science, Graduate School of Life and Environmental Sciences, University of Tsukuba, 1-1-1Tennodai, Tsukuba, Ibaraki 305-8572, Japan
e-mail: houhao880828@gmail.com

C. Zhou et al. (eds.), *Spatial Data Handling in Big Data Era*,
Advances in Geographic Information Science,
DOI 10.1007/978-981-10-4424-3_6

neighborhood environment and walking behavior (Azmi et al. 2013; Eronen et al. 2014; Van Dyck et al. 2009). Generally, walking behavior can broadly be categorized into three types: recreational, occupational and utilitarian walking. Recreational and utilitarian walking behavior are frequently compared with neighborhood environment (Saelens and Handy 2008). Recreational walking behavior refers to those undertaken in someone's leisure time without a determined destination, such as taking a walk in a park, running along the track or walking a dog. On the other hand, utilitarian walking behavior always has a specific destination and the walking is regarded as mobile means similar to riding a bicycle, taking a bus or driving a car. Considering only the physical attributes, utilitarian walking behavior tends to have a stronger relationship with neighborhood environment compared to recreational walking behavior (Lee and Moudon 2006).

Studies on evaluating neighborhood environment started with the adoption of perceived data gaining from questionnaires. One of the most widely used questionnaires is the NEWS (Neighborhood Environment Walkability Survey) developed in 2002 (Saelens et al. 2003). The questionnaire-based data is easy to be analyzed but the collecting process is both time and money consuming. As a result, this approach is not applicable to studies carried out on a large scale and most of these studies concentrated on a community level (Azmi et al. 2013; Chen et al. 2013; Kamada et al. 2009; Kondo et al. 2009). In recent years, with the development of GIS (Geographical Information System) as well as the growing number of available spatial data, studies on neighborhood environment with objective data analyzed by GIS software is becoming popular (Hanibuchi et al. 2011; Lamíquiz et al. 2015; Leslie et al. 2007). The approach based on available spatial data and GIS software can reduce the cost of collecting data. Besides, GIS software provides the function to visualize and analyze the data from the spatial view, including the capacity of mapping, spatial analysis and modeling (Leslie et al. 2007). These advantages provide a possibility to evaluate neighborhood environment on a large scale (such as a municipality level) and compare the results with the spatial patterns of urban structure and the public transportation system.

However, the adoption of GIS and objective spatial data brings several challenges in data handling. First, spatial data from different sources may differ in format, coordinate system, the definition of attributes, resolution, scale, etc. All of these differences need to be unified according to the study area. The process of unification may require simulation of some mismatched or missing data. Second, huge data often includes plenty of information. However, a specific study only needs a small part of the whole data set. As a result, the extraction of useful information (known as "data mining") is a necessary step during the data handling and this step requires knowledge of the whole data structure. Third, analyzing big data requires great computing power. Computer and software may have limitations in the maximum amount of records and the maximum data size. In this case, the data need to be divided according to the limitation of computing power and processed separately. Although these challenges exist, using GIS and objective spatial data in neighborhood environmental studies is attractive as it provides different views and understanding in this field (McGinn et al. 2007).

The purpose of this study is to evaluate the neighborhood environment and utilitarian walking behavior in Tokyo Metropolitan Area and compare the results to check the relationships. Although plenty of studies on detecting relationships between neighborhood environment and walking activity existed, limited studies were carried out in a study area covering such a big metropolitan level scale (Sundquist et al. 2015). This study is able to provide the spatial patterns of both neighborhood environment and walking behavior which are important for urban planners and public transportation designers. The produced data in this study can also be related to other social-economic data for further studies.

Methodology

This study was separated into two parts: the evaluation of neighborhood environment and the evaluation of utilitarian walking behavior. For the evaluation of neighborhood environment, we used the location data of residential buildings from Zenrin© TOWN II digital map, the road network data from OpenStreetMap Project, the land use information and spatial distribution of public transportation facilities (including the locations of bus stop and railway stations) from National Land Numerical Information constructed by the Japanese government. For the evaluation of utilitarian walking behavior, we employed the People Flow Data of Tokyo in 2008 made by CSIS (Center for Spatial Information Science), University of Tokyo.

Study Area

The Tokyo metropolitan study area is composed of the city of Tokyo, the prefectures of Chiba, Kanagawa and Saitama, and the southern part of Ibaraki prefecture (Fig. 1). The study area was decided based on the available scale of the People Flow Data. This area is known as one of the largest metropolitan areas around the world. The population of this area reached 37.6 million in 2010 and parts of the Tokyo city had the highest population densities in the world (Bagan and Yamagata 2012). The Tokyo Metropolitan Area owned the world's most extensive urban rail network. According to the latest data from the government, the public transportation system served more than 900 million passengers in 2014 (Bureau of General Affairs, Tokyo Metropolitan Government 2014).

Measures of Neighborhood Environment

Five criteria were selected to evaluate the neighborhood environments including residential density, street connectivity, land use diversity, bus stop density and

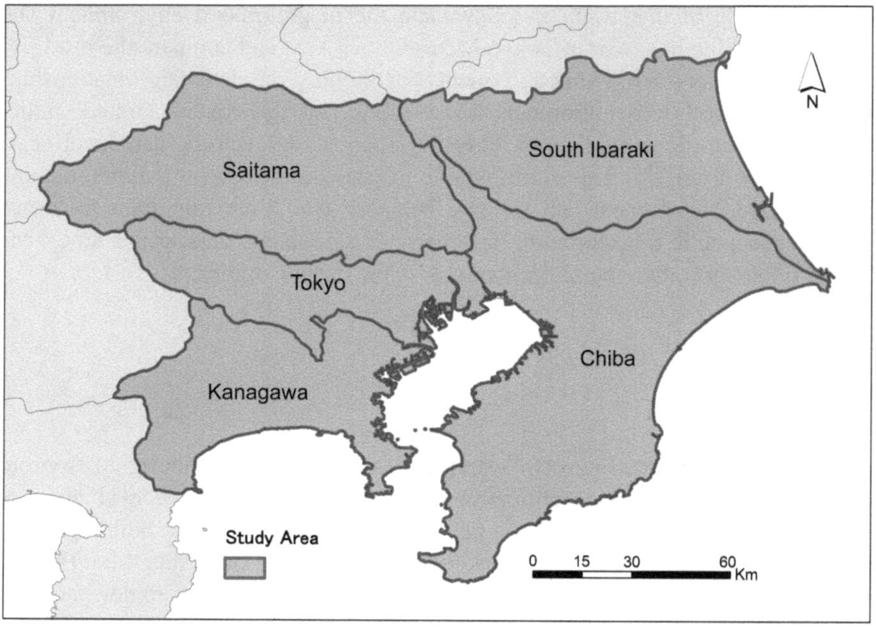

Fig. 1 Study area: Tokyo Metropolitan Area (TMA)

railway station accessibility. The first three criteria were widely used in the evaluation of neighborhood environment and walkability in the previous studies (Jun and Hur 2015; Lamíquiz and López-Domínguez 2015; Sundquist et al. 2011). The last two criteria, bus stop density and railway station accessibility, were included in this study since residents in TMA relied a lot on the public transportation in their daily lives. The neighborhood was defined as the area with a distance less than 1 km to the residence. The selection of 1 km radius buffer resulted from the evidence in previous study which showed that neighborhood environment attributes within 1 km home buffers were positively associated with moderate-vigorous physical activity in the buffer (Troped et al. 2010). Finally, through the Multi-criteria Evaluation approach, all the criteria were combined to calculate the index for evaluating neighborhood environment. The index was named "walkability", which was used to evaluate the extent to which the built environment was friendly to the presence of people's walking behavior.

Residential Density

Locations of residential buildings in TMA were derived from Zenrin© TOWN II digital maps. The first step was the combination of all the town maps. More than 200 layers were merged together with the function in the ArcGIS® software package, version 10.2. The next step was to extract residential buildings from all the

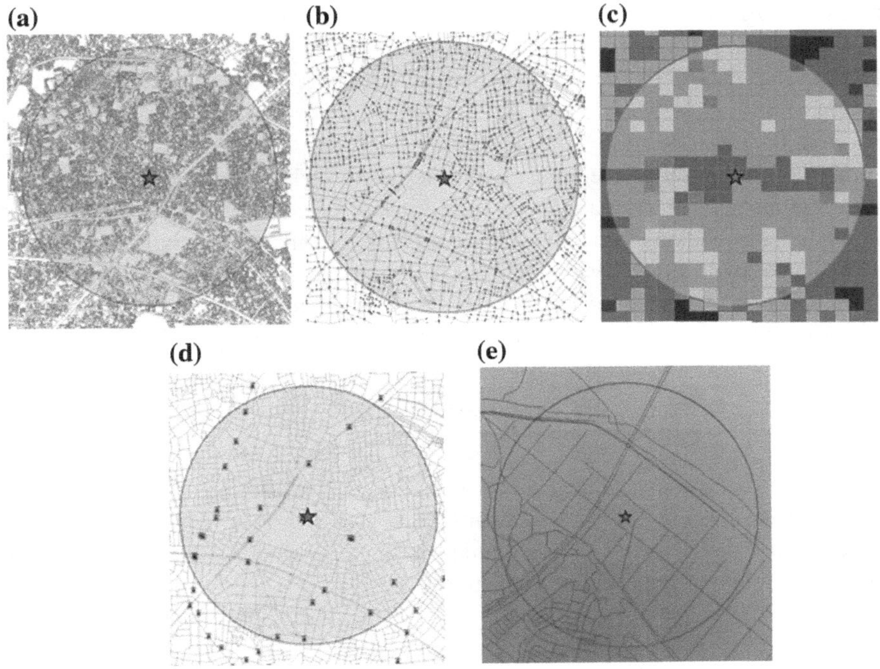

Fig. 2 Evaluation with 1-km buffer for (*left–right*): **a** residential density, **b** street connectivity, **c** land use diversity, **d** bus stop density and **e** railway station accessibility

buildings by the attribute of type. This step made the total number of features decreased from 16.4 million to 9.2 million. After this, a point-based resident's location layer obtained from the People Flow Data was added for creating the neighborhood buffers of each person. With overlay analysis, the count of residential buildings in each buffer was summarized and this value was made as the residential density of each residence (Fig. 2a).

Street Connectivity

In this study, the street connectivity was evaluated by the number of intersections within each neighborhood. Data from OpenStreetMap Project were utilized to get the road layer. Later, according to the description of the road categories, only the roads available for walking behavior were extracted. Next, the "network analysis" function, which is available in the ArcGIS® software package, version 10.2, was used to building road network and get intersections. Finally, the layer of neighborhood buffers created before were overlaid with the layer of intersections to get the count of intersections within each neighborhood as the value of street connectivity (Fig. 2b).

Land Use Diversity

The original data used for the measurement of land use diversity came from the 100 m × 100 m land use mesh data included in the National Land Numerical Information constructed by the Japanese government. The original data had a number of 12 land use categories. Later they were reclassified into five categories since the purpose of evaluating this factor was to detect potential destinations for people's daily walking behavior. The five categories included: single-family residential area, multifamily residential area, commercial area, public service area and green space. Land use diversity was calculated by the formula below and the value (d) represented the diversity of each person's neighborhood (Fig. 2c):

$$d = \mathbf{I}\frac{\sum k(p_k \ln p_k)}{\ln N}$$

where d is the diversity value; k is the category of land use; p is the proportion of each land use category; N is the number of land use categories. The equation results in between 0 and 1, with 0 representing a single type of all land use and 1, a developed area with all land use categories.

Bus Stop Density

The bus stop density value was defined as the count of bus stop in each neighborhood buffer (Fig. 2d). The original data recording the spatial location of bus stop were derived from the National Land Numerical Information. The number of bus stop in each resident's neighborhood indicated the scale of accessible areas reached by taking a bus. With a higher bus stop density, residents in the neighborhood intended to have a higher possibility to choose bus as the movement means. When people choose to go out by bus, the utilitarian walking behavior usually happens since they need to take a walk to reach the bus stops.

Railway Stations Accessibility

The railway station accessibility was evaluated through the Euclidean Distance from each residential point to the closest railway station. The raster layer with a cell size of 100 m was created and the value of each cell was the distance to the nearest railway station. The neighborhood buffers were later utilized to get the average value of distance in each neighborhood (Fig. 2e). As mentioned above, good access to the public transportation facilities can encourage the utilitarian walking behavior with the purpose of reaching those facilities.

Multi-criteria Evaluation Approach

After the evaluation of five criteria, equal weights were given to the value of each criterion to calculate the final walkability. The decision of weights was based on the previous study (Leslie et al. 2007). All the values were normalized to force the values to fall into 0 and 1. As a result, the final values of walkability ranged between 0 and 5. According to the values of walkability, the whole areas were categorized into five groups shown in Table 1.

Measures of Utilitarian Walking Behavior

In this study, the total utilitarian walking time per day was adopted as the value to present each person's level of utilitarian walking behavior. The data source for the measures was the People Flow Data. The People Flow Data is a data set processed for monitoring dynamic changes in daily people flow, which provides the individual locations in every minute within 24 h. The procedures of data processing included: (a) geocoding the first and last points of sub-trips to specify spatiotemporal locations, (b) calculating the shortest route between the two locations, and (c) interpolating minute-to-minute location information based on detailed network data. In this study, the People Flow Data of Tokyo in 2008 was used for the measurement and the total number of samples reached 576,806. Table 2 showed the structure of the People Flow Data. The critical fields used in this study are PID, LON, LAT, PURPOSE, and TCODE. TCODE helped to extract only the walking behavior. The spatial information of the walking activities was recorded by LON and LAT. PURPOSE was for extracting only the utilitarian walking from all the walking behavior. Table 3 shows all the purpose of walking behavior. In this study, the authors tried to ignore those occasional walking behavior and to focus on only the utilitarian walking behavior happened almost every day. In this context, only the first four (code 1–4) categories were considered as the utilitarian walking behavior. After the extraction, the records were summarized based on PID to link the walking behavior with the neighborhood environment.

Table 1 Area separation based on walkability

Walkability (0 ~ 5)	Category
0–1	Low walkable area
1–2	Medium low walkable area
2–3	Medium walkable area
3–4	Medium high walkable area
4–5	High walkable area

Table 2 Structure of people flow data

Field ID	Field name	Description
1	PID	Unique person ID
2	TNO	Trip number
3	SNO	Sub trip number
4	LON	Longitude position
5	LAT	Latitude position
6	GENDER	Gender
7	AGE	Age group
8	ZCODE	Current location by zone code
9	OCCUP	Person occupation
10	PURPOSE	Purpose to trip
11	MAGFAC	Adjustment factor
12	MAGFAC2	Adjustment factor
13	TCODE	Mode of transportation

Table 3 Purpose code in people flow data

Code	Value	Code	Value
1	To office	9	To send/pick up activity
2	To school	10	For selling and buying
3	To home	11	For appointment
4	To shopping place	12	To/for work (fixing and repairing)
5	For dinner/short recreation	13	To agriculture/forestry/fishery work
6	For sightseeing and leisure	14	Other business purpose
7	For medical treatment	99	Others
8	For other private purpose		

Visualization with Standard 1 Km × 1 Km Mesh

After the evaluation process, all the results were summarized by the standard 1 km × 1 km grid net established by the Geospatial Information Authority of Japan. The value of each grid was determined by the average value of all the residential points that fell into this grid. There were two objectives for this approach: creating standard data and visualization. Data summarized by the standard 1 km × 1 km grid net is applicable for comparative studies with other social and economic data published by the Japanese government which utilized the same unit. Instead of point-based results, the grid-based results are clearer for visualization and easier for detecting the spatial patterns of the results.

Results

Evaluation Results of Five Criteria

Figure 3 showed the evaluation results for the five criteria separately. Results of residential density (Fig. 3a) showed that except for the Chu'o ward which was located in the central of TMA, the residents in the other 22 special wards of Tokyo all had a high residential density. Besides these areas, the high residential area appeared along the railway lines revealed a common pattern that people intended to live in places with a good accessibility to the railway stations. The low value appeared in both the central area of Tokyo and the rural areas of the metropolitan area. Low residential density in the central area resulted from that most of the buildings there were commercial land use. On the other hand, rural areas had a low residential density because of the low population density there. Street connectivity showed a similar spatial pattern with the residential density (Fig. 3b) that the highest value appeared in the urban areas with a short distance to the urban core while the lowest values appeared in the rural areas far from the urban core. The spatial patterns of these two criteria can be understood from the perspective of urban structure. The suburban areas close to the city center were usually designed as the residential areas with a high density of residential buildings and standard road networks. The result of land use diversity (Fig. 3c) had a slim difference compared with the first two criteria. Although the lowest value was still assigned to the rural areas, the highest value appeared both in the urban core and the urban areas relatively close to the urban core. The diverse land use in the central area resulted from the need to serve the big flowing population passed there every day. What's more, some suburban areas had the same low values as rural areas. This indicates that some of the residential areas in Tokyo might be in a lack of enough facilities for daily life in the neighborhood context. Results of the bus stop density (Fig. 3d) and the railway station accessibility (Fig. 3e) showed similar spatial patterns. The areas within the urban boundary had higher values than those rural areas. This proved that the Tokyo city had a complete public transportation system to serve all the citizens regardless of the distance to the city center while in rural areas only residents living in places close to the railway lines enjoyed good accessibility to the public transportation facilities.

Evaluation Results of Neighborhood Environment (Walkability)

The five criteria were merged together with the equal weight and the result was shown in Fig. 4. All the areas were categorized into five groups (Table 1) by the value of their walkability. Most of the high walkable areas concentrated in the 23 special wards of Tokyo except the Chu'o ward. Residents here enjoyed a good

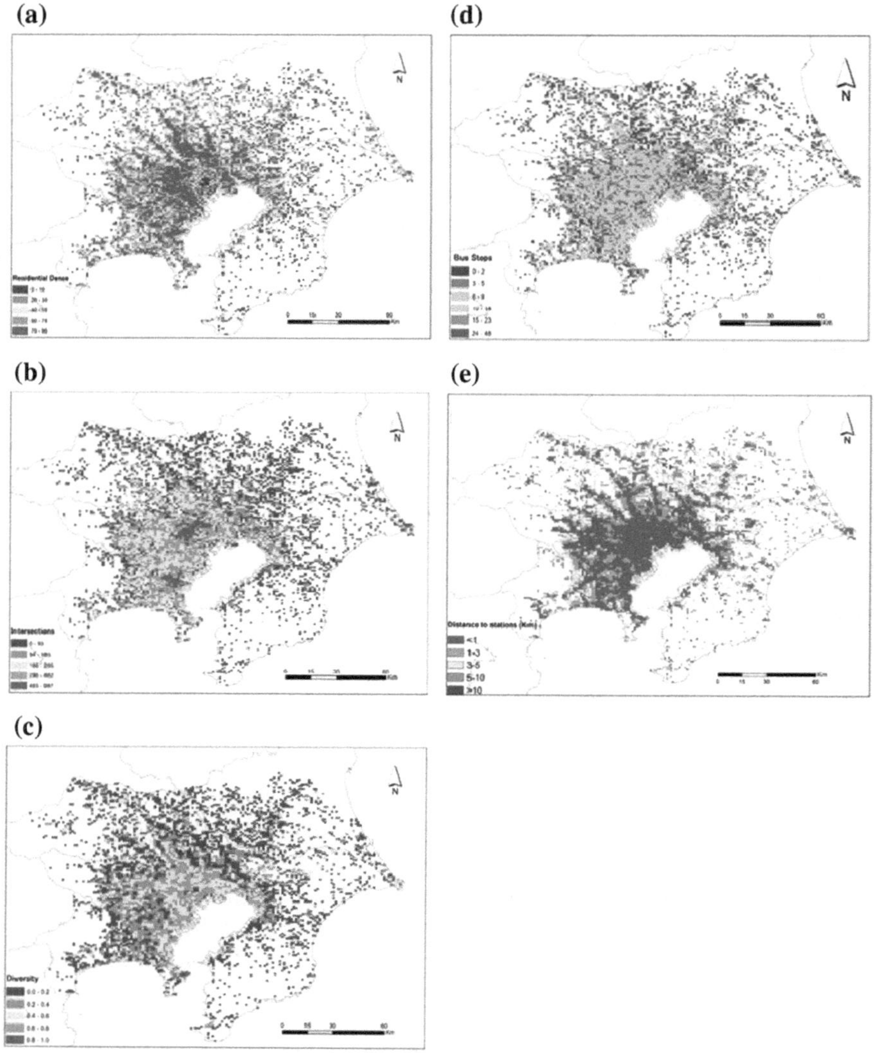

Fig. 3 Grid-based maps of **a** residential density; **b** street connectivity; **c** land use diversity; **d** bus stop density; **e** railway station accessibility

accessibility to public transportation facilities which encourages them to have a walk to reach stations. The high diversity of land use here provided plenty of potential destinations for residents to walk to within the neighborhood scale. The complex road network here reduced the potential to move by a private car. The medium walkable area appeared along the railway lines as well as the municipal lines between Special wards of Tokyo and other prefectures. Residents here owned a good accessibility to the public transportation facilities and the residential buildings. However, the diversity of land use was relatively low compared to the high walkable areas, which indicated a low potential for daily walking behavior within the neighborhood. Low walkable areas scattered in the rural areas with the longest distance to the city center compared to the other categories. Residents here suffered a bad accessibility to the public transportation facilities, and it led to a high potential to use a private car for daily movement. The low residential density and land use diversity here reduced the chance for residents to reach a destination by walking since the potential destinations were far from their living places. The walkability map was related to the urban structure from the spatial perspective that except for the central business district (the Chu'o ward), the walkability decreased when the distance to the urban core increased.

Evaluation Results of Utilitarian Walking Behavior

Results of the utilitarian walking time (Fig. 5) showed that rural residents' utilitarian walking time per day were less than the time of people who live in the urban core and suburban areas. Most of the residents in rural areas had a utilitarian walking time of fewer than 10 min per day. While residents in the suburban areas close to the boundary of each prefecture usually walked more than 10 min per day for the utilitarian purposes. But most of them didn't reach the 30 min utilitarian walking time in one day. People living in the urban areas close to the city center had

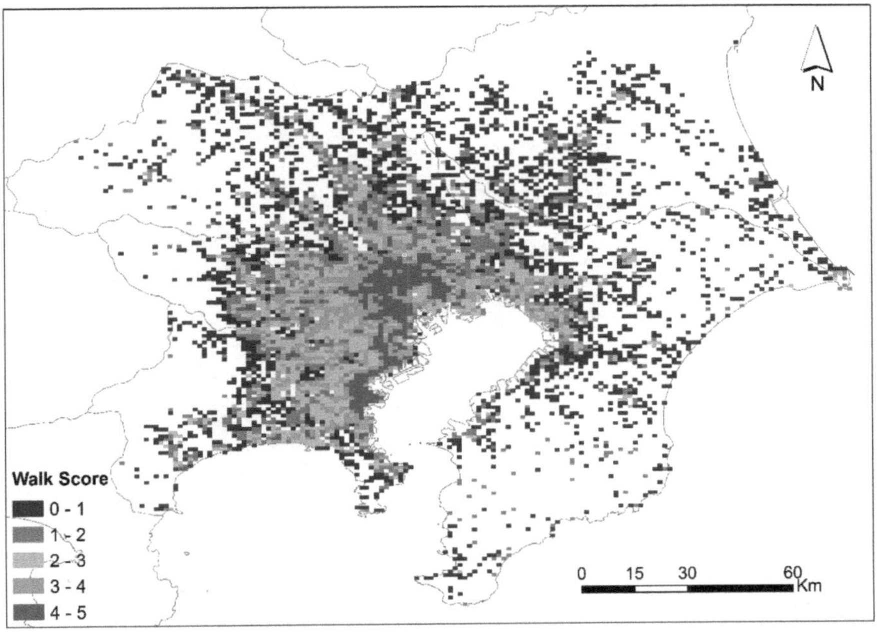

Fig. 4 Grid-based map of walkability in TMA

a higher average utilitarian walking time per day. This result kept a consistence with the findings from the evaluation of walkability. The majority of the residents in this area had a utilitarian walking time reaching the level of 20–30 min. Residents with more than 30 min utilitarian walking time per day could be easily found in this area. The utilitarian walking behavior of residents in the Chu'o ward was slimly mismatched with the findings of walkability evaluation. Residents here had a similar level of daily utilitarian walking behavior with the residents living close to the city center (mostly in 20–30 min level and 30–40 min level) although the walkability in Chu'o ward was less than the surrounding areas.

Comparison Between Walkability and Utilitarian Walking Time

By comparing Figs. 4 and 5, similar spatial patterns can be detected that residents in the rural areas have low walkability in the neighborhood and low utilitarian walking time. On the other hand, residents in the urban areas, especially areas close to the city center, enjoyed high walkability and had more utilitarian walking time per day. With the statistics shown in Table 4, the consistence between the evaluation results of walkability and utilitarian walking time was clearer. Although the maximum utilitarian walking time shown in the second column did not match the walkability

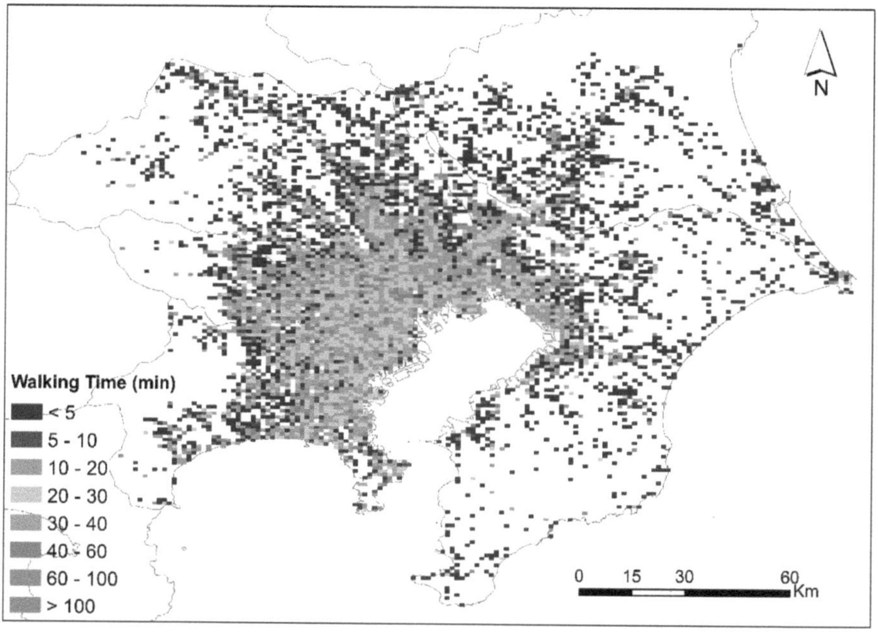

Fig. 5 Grid-based map of utilitarian walking time in TMA

Table 4 Comparison of walkability and utilitarian walking time

Walkability	Max UWT (min)	Mean UWT (min)	Count of grids	Proportion (%)
0–1	152	9.6	1288	22.38
1–2	144	11.7	1560	27.11
2–3	182	17.1	1410	24.50
3–4	62	22.2	961	16.70
4–5	82	24.9	535	9.30

value, the mean utilitarian walking time perfectly matched the walkability. Residents who live in the areas assigned with a low walkability value do have lower mean utilitarian walking time every day. Count of grids showed the proportion of each category valued by walkability in TMA. The table showed that the medium low walkable area (value: 1–2) covered the largest area of TMA, followed by the medium walkable area (value: 2–3) and the low walkable area (value: 0–1). The left two categories, the medium high and the high walkable areas, only covered 26% of the whole study area.

Discussion

The evaluation of neighborhood environment for walking is popular in recent years as people pay more attention to personal health. Many studies tried to evaluate neighborhood environment or walkability with both objective and self-reported data. However, because of the differences in the study area and personal attributes, there is no standard approach for all the analysis. The differences mainly appear in the selection of criteria considering the purpose of each study. For example, one study about detecting the effect of neighborhood environment on walking for transportation adopted street connectivity, land use mix and residential density as the criteria for evaluation (Turrell et al. 2013). Another study aiming at finding the association between destination and route attributes with walking chose sidewalks, street connectivity, aesthetics, traffic and safety as the criteria (Sugiyama et al. 2012). As a result, when doing researches on this field, there is a need to consider about the selection of which criteria should be included. Studies of neighborhood environment carried out in America or Europe usually don't use public transportation factors (Sundquist et al. 2011; Troped et al. 2010) while researchers doing studies in Japan need to consider this factor as public transportation system is widely and frequently used here.

The main purpose of this study is to detect the relationship between neighborhood environment and utilitarian walking behavior. Although more detailed and deeper statistical analysis was needed, the results reflected that people living in high walkable areas really had more average utilitarian walking time. Studies focusing on utilitarian walking behavior were still limited in this field. Previous studies only proved that moving to a more walkable neighborhood was associated with an increase in utilitarian walking time (Hirsch et al. 2014). And utilitarian walking behavior had a positive association with the local accessibility to amenities (Wasfi et al. 2015). Since utilitarian walking behavior is the most common walking behavior happened almost every day, more studies are needed to detect the potential ways to improve the level of this walking behavior. Besides the findings of relationship, this study also released the maps of five neighborhood attributes, walkability, and utilitarian walking time. All the maps showed the spatial patterns similar to the urban structure. Previous studies mostly concentrated on a micro scale, but the findings here showed a possibility of comparing the neighborhood environment in the whole urban structure.

The GIS-based objective measurement for neighborhood environment walkability seems to be more reliable than the perceived subjective measurements if the accuracy of the spatial data is acceptable because participants' perception of their neighborhood may vary even if they live in the same place. With the increasing computing capabilities, the GIS-based objective measurement provides a considerable opportunity to develop more accurate measures of the neighborhood environment. This study showed one basic way of interpreting related spatial data together for the evaluation and it also proved that GIS is suitable for handling big spatial data. With the technical developments in computer science and the increase of available open data sources, the GIS-based objective measurement is supposed to behave better in the future.

There are several limitations to this study. First is the missing of some potential variables that may improve the results. Although the consistence can be found by comparing the evaluation results of walkability and utilitarian walking time, some areas such as the Chu'o ward showed mismatched patterns. Further study is needed for adding new variables to check and improve the results. Second is the use of self-reported data for the evaluation of utilitarian walking time. The People Flow Data is excellent since it covers the whole Tokyo Metropolitan Area with a big number of sample. However, the original data came from the self-reported questionnaires and this kind of data is hard to remove the influence of subjective bias to recall and response (Kamada et al. 2009). Third is the ignorance of personal attributes. This study utilized the whole data set without extraction of any specific groups of people. However, previous studies have proved that people's walking behavior were related to personal attributes such as age (Hanibuchi et al. 2011), gender (Van Dyck et al. 2013), income (Owen et al. 2007), driving status (Kamada et al. 2009), etc. In order to increase the accuracy of the results, the extraction of different groups was considered in the future study.

Conclusion

The results showed that residents in urban areas with a good accessibility to the city center had the highest potential for daily utilitarian walking behavior, followed by the residents in the urban core and rural areas. The spatial patterns of the result had a consistence with the result of personal utilitarian walking time derived from the People Flow Data. This consistence proved that residential density, street connectivity, land use diversity, bus stop density, railway station accessibility are necessary factors for evaluating neighborhood environment in TMA. The evaluation of neighborhood environment reflected the reality and the results can be utilized by both urban planners and transportation network designers for building a more walkable city. Future studies are encouraged on deeper statistical analysis of the relationships between neighborhood environment and utilitarian walking time to increase the confidence of the findings.

This study employed an 1-km radius buffer to calculate each criterion to evaluate neighborhood environment for daily walking behavior and basic grids for visualizing the spatial patterns of the evaluation results. The outcomes of this study supported that the handling of spatial data about neighborhood environment with certain buffers was reasonable. The approach of processing objective GIS data and subjective questionnaire-based data in this study was worth to be applied to other metropolitan areas.

References

Azmi, D. I., Karim, H. A., & Ahmad, P. (2013). Comparative study of neighbourhood walkability to community facilities between two precincts in Putrajaya. *Procedia-Social and Behavioral Sciences, 105,* 513–524.

Bagan, H., & Yamagata, Y. (2012). Landsat analysis of urban growth: How Tokyo became the world's largest megacity during the last 40 years. *Remote Sensing of Environment, 127,* 210–222.

Bureau of General Affairs, Tokyo Metropolitan Government. (2014). Tokyo Statistical Yearbook 2014: Transport. http://www.toukei.metro.tokyo.jp/tnenkan/2014/tn14q3e004.htm. Accessed June 12, 2016.

Chen, T. A., Lee, J. S., Kawakubo, K., Watanabe, E., Mori, K., Kitaike, T., et al. (2013). Features of perceived neighborhood environment associated with daily walking time or habitual exercise: Differences across gender, age, and employment status in a community–dwelling population of Japan. *Environmental Health and Preventive Medicine, 18*(5), 368–376.

Eronen, J., von Bonsdorff, M. B., Törmäkangas, T., Rantakokko, M., Portegijs, E., Viljanen, A., et al. (2014). Barriers to outdoor physical activity and unmet physical activity need in older adults. *Preventive Medicine, 67,* 106–111.

Hallal, P. C., Andersen, L. B., Bull, F. C., Guthold, R., Haskell, W., & Ekelund, U. (2012). Lancet physical activity series working group. (2012). Global physical activity levels: surveillance progress, pitfalls, and prospects. *The Lancet, 380*(9838), 247–257.

Hanibuchi, T., Kawachi, I., Nakaya, T., Hirai, H., & Kondo, K. (2011). Neighborhood built environment and physical activity of Japanese older adults: results from the Aichi Gerontological Evaluation Study (AGES). *BMC Public Health, 11*(1), 1.

Heath, G. W., Parra, D. C., Sarmiento, O. L., Andersen, L. B., Owen, N., Goenka., S., …, Lancet Physical Activity Series Working Group. (2012). Evidence-based intervention in physical activity: Lessons from around the world. *The Lancet, 380*(9838), 272–281.

Hirsch, J. A., Diez Roux, A. V., Moore, K. A., Evenson, K. R., & Rodriguez, D. A. (2014). Change in walking and body mass index following residential relocation: The multi-ethnic study of atherosclerosis. *American Journal of Public Health, 104*(3), e49–e56.

Inoue, S., Ohya, Y., Tudor-Locke, C., Tanaka, S., Yoshiike, N., & Shimomitsu, T. (2011). Time trends for step-determined physical activity among Japanese adults. *Medicine and Science in Sports and Exercise, 43*(10), 1913–1919.

Jun, H. J., & Hur, M. (2015). The relationship between walkability and neighborhood social environment: The importance of physical and perceived walkability. *Applied Geography, 62,* 115–124.

Kamada, M., Kitayuguchi, J., Inoue, S., Kamioka, H., Mutoh, Y., & Shiwaku, K. (2009). Environmental correlates of physical activity in driving and non-driving rural Japanese women. *Preventive Medicine, 49*(6), 490–496.

Kondo, K., Lee, J. S., Kawakubo, K., Kataoka, Y., Asami, Y., Mori, K., et al. (2009). Association between daily physical activity and neighborhood environments. *Environmental Health and Preventive Medicine, 14*(3), 196–206.

Lamíquiz, P. J., & López-Domínguez, J. (2015). Effects of built environment on walking at the neighbourhood scale. A new role for street networks by modelling their configurational accessibility? *Transportation Research Part A: Policy and Practice, 74,* 148–163.

Lee, C., & Moudon, A. V. (2006). Correlates of walking for transportation or recreation purposes. *Journal of Physical Activity & Health, 3,* S77.

Leslie, E., Coffee, N., Frank, L., Owen, N., Bauman, A., & Hugo, G. (2007). Walkability of local communities: Using geographic information systems to objectively assess relevant environmental attributes. *Health & Place, 13*(1), 111–122.

McGinn, A. P., Evenson, K. R., Herring, A. H., Huston, S. L., & Rodriguez, D. A. (2007). Exploring associations between physical activity and perceived and objective measures of the built environment. *Journal of Urban Health, 84*(2), 162–184.

Ng, S. W., Norton, E. C., & Popkin, B. M. (2009). Why have physical activity levels declined among Chinese adults? Findings from the 1991–2006 China Health and Nutrition Surveys. *Social Science and Medicine, 68*(7), 1305–1314.

Owen, N., Cerin, E., Leslie, E., Coffee, N., Frank, L. D., Bauman, A. E., et al. (2007). Neighborhood walkability and the walking behavior of Australian adults. *American Journal of Preventive Medicine, 33*(5), 387–395.

Saelens, B. E., & Handy, S. L. (2008). Built environment correlates of walking: A review. *Medicine and Science in Sports and Exercise, 40*(7 Suppl), S550.

Saelens, B. E., Sallis, J. F., Black, J. B., & Chen, D. (2003). Neighborhood-based differences in physical activity: An environment scale evaluation. *American Journal of Public Health, 93,* 1552–1558.

Sugiyama, T., Neuhaus, M., Cole, R., Giles-Corti, B., & Owen, N. (2012). Destination and route attributes associated with adults' walking: A review. *Medicine and Science in Sports and Exercise, 44*(7), 1275–1286.

Sundquist, K., Eriksson, U., Kawakami, N., Skog, L., Ohlsson, H., & Arvidsson, D. (2011). Neighborhood walkability, physical activity, and walking behavior: The Swedish Neighborhood and Physical Activity (SNAP) study. *Social Science and Medicine, 72*(8), 1266–1273.

Sundquist, K., Eriksson, U., Mezuk, B., & Ohlsson, H. (2015). Neighborhood walkability, deprivation and incidence of type 2 diabetes: A population-based study on 512,061 Swedish adults. *Health & Place, 31,* 24–30.

Troped, P. J., Wilson, J. S., Matthews, C. E., Cromley, E. K., & Melly, S. J. (2010). The built environment and location-based physical activity. *American Journal of Preventive Medicine, 38*(4), 429–438.

Turrell, G., Haynes, M., Wilson, L. A., & Giles-Corti, B. (2013). Can the built environment reduce health inequalities? A study of neighbourhood socioeconomic disadvantage and walking for transport. *Health & Place, 19,* 89–98.

Van Dyck, D., Cerin, E., Conway, T. L., De Bourdeaudhuij, I., Owen, N., Kerr, J., et al. (2013). Perceived neighborhood environmental attributes associated with adults' leisure-time physical activity: Findings from Belgium, Australia and the USA. *Health & Place, 19,* 59–68.

Van Dyck, D., Deforche, B., Cardon, G., & De Bourdeaudhuij, I. (2009). Neighbourhood walkability and its particular importance for adults with a preference for passive transport. *Health & Place, 15*(2), 496–504.

Wang, Y., Chau, C. K., Ng, W. Y., & Leung, T. M. (2016). A review on the effects of physical built environment attributes on enhancing walking and cycling activity levels within residential neighborhoods. *Cities, 50,* 1–15.

Wasfi, R. A., Dasgupta, K., Eluru, N., & Ross, N. A. (2015). Exposure to walkable neighbourhoods in urban areas increases utilitarian walking: Longitudinal study of Canadians. *Journal of Transport & Health* (in press).

A Space-Time GIS for Visualizing and Analyzing Clusters in Large Tracking Datasets

Hongbo Yu

Introduction

Large tracking datasets of moving objects are becoming increasingly available in various research fields due to recent advancements of information and location-aware technologies (Laube et al. 2007). Moving objects in general refer to objects whose location and/or shape may change over time (Erwig et al. 1999). In many tracking datasets, the concerned moving objects usually maintain their shape and identity while their locations change over time (Dodge et al. 2009). These moving objects, such as individual people, vehicles, and wild animals, are recognized as moving points (Erwig et al. 1999). Their movements then can be represented as trajectory lines. A tracking dataset of moving objects contains many such trajectories, which record the locations where each moving object visited and when the object was there. Such datasets provide a unique information source for researchers to explore spatiotemporal distribution of the observed objects. Information derived from such datasets can help researchers identify the locations where and when many observed objects cluster together. Such locations could be places where a group of people gather together for specific activities, bottleneck locations in a transportation network where traffic congestion occurs, or habitat areas of wild animals where they live or hunt for food. Being able to identify these locations and understand their spatiotemporal characteristics can contribute to the knowledge base of the related research areas, especially where such locations have not been clearly identified or well understood. Therefore, large tracking datasets provide promising opportunities for researchers to discover these key locations

H. Yu (✉)
Department of Geography, Oklahoma State University, Stillwater, OK 74078, USA
e-mail: hongbo.yu@okstate.edu

C. Zhou et al. (eds.), *Spatial Data Handling in Big Data Era*,
Advances in Geographic Information Science,
DOI 10.1007/978-981-10-4424-3_7

related to the observed moving objects. However, when a large number of objects are involved, it becomes very difficult to identify these locations and discern their spatiotemporal patterns (Dodge et al. 2009). Thus, analysis tools are needed to effectively represent the dataset, restructure the data, and derive useful information (Purves et al. 2014).

Time geography (Hägerstrand 1970), which supports an integrated space-time system for examining relationships between individual's activities and their spatiotemporal constraints, has a natural fit for representing individual-based tracking data and thus provides an elegant approach to studying individual's movements. However, limited geographic computational power in the past has constrained the development of an operational system of the time-geographic framework (Yuan et al. 2004). Recent advancements in computational technology have significantly increased the capability of geographic information system (GIS) to represent, process, and analyze spatial data. Hence, GIS has been suggested as a useful platform to support the implementation of the time-geographic framework and facilitate the management and analysis of tracking datasets. A number of early efforts (e.g., Miller 1991; Kwan and Hong 1998; Kwan 2000a) demonstrated the possibility of implementing the key concepts of time geography in a two-dimensional (2D) GIS environment. However, lacking an integrated time dimension in the design, current mainstream GIS falls short of providing an effective environment to handle tracking datasets which contain rich spatiotemporal information. The current GIS design needs to be extended to support the representation and manipulation of trajectories of moving objects. As there is a revived interest in time geography in the research community, a number of recent attempts have explored the possibility of using a three-dimensional (3D) GIS environment to simulate the space-time system of time geography and provide 3D visualization of space-time paths and prisms in GIS (Kwan 2000b; Buliung and Kanaroglou 2006; Yu 2006; Andrienko et al. 2007; Neutens et al. 2008; Yu and Shaw 2008; Shaw et al. 2008; Shaw and Yu 2009; Kveladze et al. 2015). These studies confirm the possibility of using a 3D GIS environment to operationalize the time-geographic framework, showcase the advantages of interactively visualizing time-geography concepts (e.g., space-time paths and prisms), and demonstrate the potential of using such an analysis environment to support the exploration of spatiotemporal relationships among moving objects. However, it remains a great research challenge to apply such an approach to large tracking datasets. Extending from the existing approaches, this study attempts to develop analysis approaches in a space-time GIS environment to help researchers investigate trajectories stored in large tracking datasets and explore the locations where the paths of the objects cluster in space and time.

The rest of this paper is organized into four sections. The next section includes discussions on related research topics, including moving objects, time geography and space-time GIS. Section 3 introduces the station concept and discusses how this concept can be used to help researchers investigate the spatial and temporal characteristics of places where the paths of objects converge. Several spatial and temporal aggregation methods are proposed to explore the spatial and temporal extent of stations presented in moving objects datasets. A space-time GIS framework that can support the representation of trajectories and stations is introduced in Sect. 4 and some station analysis results based on a sample individual-based tracking dataset are reported. Finally, concluding remarks are provided in Sect. 5.

Related Research

Many objects in the world move across the space, with or without the change of their shapes. In the literature of spatiotemporal objects, such objects are identified as moving objects (Erwig et al. 1999; Laube et al. 2007; Dodge et al. 2009). Representing a common yet simple case, many moving objects change their locations with fixed shape and identity. In this case, moving objects are recognized as moving points and may be represented as point features. In many tracking datasets of moving objects, shape change does not happen or is not of concern. Therefore, a moving object's trajectory, which records the movement history of the object, can be constructed from a sequence of time-stamped point locations visited by the object. Rich spatiotemporal information of an object's movements is embedded in such a trajectory. However, when the number of objects in a tracking dataset increases, their trajectories may become so tangled and it can be a challenging task to discern anything meaningful from such a dataset. Many studies have attempted to untangle the twisted trajectories and reveal useful information hidden in the lines. A common approach is to develop certain movement descriptors which can be used to capture the characteristics of the object's movement trajectory and simplify the representation of the trajectories (Laube et al. 2007; Dodge et al. 2009). The descriptors can be for the overall shape of a trajectory (e.g., total distance, movement duration, average speed, straightness) or a range of movement properties along a trajectory (e.g., velocity, acceleration, moving direction). These descriptors then can be used to analyze similarities in movement behavior among the trajectories and identify certain movement patterns presented in the trajectory dataset (Laube and Purves 2006; Laube et al. 2007; Dodge et al. 2009; Long and Nelson 2013; Postlethwaite et al. 2013). In these approaches, individual trajectory usually is the focus for deriving the descriptors and the temporal property of a trajectory is either ignored, converted to a duration measure, or treated in a relative time manner. However, when investigating the spatiotemporal characteristics of places where the moving objects gather, it is important and necessary to deal with the original temporal property of the trajectories and analyze the spatiotemporal relationships among the trajectories. A system that embraces an integrated spatial and temporal representation therefore is needed to effectively model and analyze the trajectories to support the investigation.

Hägerstrand (1970) introduced time geography to study human activities and their constraints in a space-time context. In recent years, this framework has drawn great interests among researchers to study the trajectories of moving objects. Time geography adopts a three-dimensional (3D) orthogonal system, with two dimensions for space and one dimension for time, to study individual's movements in space and time. The space-time path concept of time geography can be readily used to model the trajectories of moving objects (Long and Nelson 2013). Represented as a linear feature in the space-time system, a space-time path allows a continuous representation of the history of an object's changing positions. In general, a path can be constructed from a sequence of two types of segments: vertical segments and tilted segments. While a vertical segment represents an object's stay at a specific location, a tilted segment records an object's movements between two places.

The space-time system of time geography also offers an effective environment for analyzing various spatiotemporal relationships (e.g., co-location in time, co-location in space, and co-existence) among the trajectories when they are represented as a set of space-time paths (Parkes and Thrift 1980; Golledge and Stimson 1997). Among many defined spatiotemporal relationships of paths, the co-existence relationship, which exists when many paths reach and stay at the same location during the same time period, requires constraints in both space and time. Identifying locations where paths co-exist usually plays an important role in investigating spatiotemporal distribution of the observed objects (Yu 2006; Andrienko et al. 2007). In time geography, the concept of station has been used to describe a location where paths cluster in space and time (Pred 1977; Gollege and Stimson 1997; Miller 2005). A station is defined as a place where people can gather and participate in activities. At a station, many individuals will share some time together and their space-time paths will form a co-existence relationship. In the 3D space-time system, a station can be recognized at a place where the vertical segments of multiple paths bundle at a specific location and stay for a certain period of time (see Fig. 1). A tube is usually used to represent the existence of a station and describe its extent in the space-time system. The spatial and temporal extent of a station may vary significantly in different applications (Gollege and Stimson 1997). A station can be a building, a city, or a region in space and its lifespan can range from a couple of hours to decades or even longer. The tubes, which confine the bundled paths in the 3D space-time system, can effectively represent stations and portray their spatial and temporal extents. Therefore, the station concept provides an

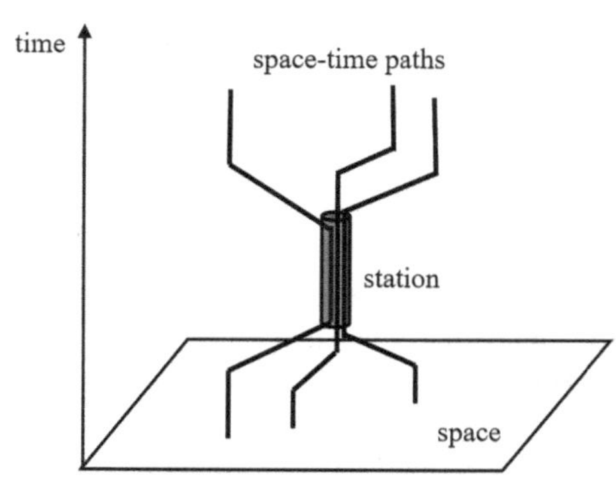

Fig. 1 Space-time paths and station

effective guidance for identifying where and when a large amount of space-time paths converge. With the 3D space-time system and the concepts of space-time path and station, time geography offers a useful theoretical foundation for exploring important locations where moving objects cluster in space and time.

Existing studies have shown the potential of GIS in managing moving objects datasets (e.g., Wolfson et al. 1998; Porkaew et al. 2001; Vazirgiannis and Wolfson 2001; Brinkhoff 2002; Dykes and Mountain 2003). These studies attempt to manage moving objects and their trajectories in a two-dimensional (2D) GIS framework, storing the temporal information of the observed objects as a non-spatial attribute in the table associated with the geographic layer. Such an approach works well for certain tasks such as maintaining the dataset and searching for records. However, without an integrated space-time system, the current 2D GIS framework cannot effectively model the rich spatiotemporal information stored in a large tracking dataset. Recent efforts have implemented the space-time system of time geography in GIS and developed a space-time GIS to support the visualization and analysis of space-time paths (see Güting et al. 2000; Kwan 2000b; Buliung and Kanaroglou 2006; Yu 2006; Shaw et al. 2008; Shaw and Yu 2009). Simulating the space-time system, these studies adopt a 3D GIS environment (2D space + 1D time) to support the representation, visualization, and analysis of paths. The space-time GIS environment allows an implementation of space-time path in a more straightforward manner to its original format. Moreover, it opens up further opportunities to operationalize other time geography concepts such as stations and support advanced spatiotemporal analysis applied to space-time paths.

A 3D space-time GIS also presents possibility to represent and visualize the station concept of time geography. In the classic time geography literature, a shape of tube has been used to describe the spatial and temporal extent of a station (Pred 1977; Gollege and Stimson 1997). Such tubes can be represented as space-time cylinders in the 3D space-time GIS. Several studies have attempted to implement the space-time cylinder approach for different types of spatiotemporal datasets. Kulldoff (2001) describes a space-time cylinder around a centroid of a census area as a geographic surveillance method for monitoring time periodic diseases. The height of the cylinder increases with increasing time and the width is determined by the radius based on the population at risk in the census area. The cylinder however is implemented by statistical methods and lacks an interactive visualization environment for spatiotemporal pattern detection and recognition. Onozuka and Hagihara (2007) employ a spatial scan statistic technique and use 3D cylindrical windows to study the geographic distribution and prediction of tuberculosis clusters in Japan. The base of the cylinder represents spatial extent and its height represents time. Both the spatial base and starting time for the cylinder are flexible and mutually independent. The methodology however does not capture fixed spatial units for which a centroid can be used as representative location. Rinner (2004) introduces a tool to model and visualize basic time geography concepts for exploring activity-travel patterns. Simple cylinder shapes with growing or shrinking cross-sections are used to represent and visualize stations. Even though the sizes of cylinders are not directly related to the number of paths clustering at the locations, they do provide an effective visualization of the station concept.

The existing literature indicates that a space-time GIS approach can be an effective and promising approach to examining clusters of trajectories. Building upon the existing space-time GIS design, this study will develop spatiotemporal analysis tools to facilitate the exploration of stations presented in large tracking datasets and support the spatiotemporal visualization of these stations in a GIS environment.

An Aggregation Approach to Deriving Stations from Space-Time Paths

In the past several decades, we have witnessed the growth of GIS in its capability of managing and representing spatial data. Since the interactive mapping environment of GIS allows more convenient visualization of geographic phenomena and their spatial relationship, it is not surprising that GIS have been frequently used to support geovisualization and exploratory data analysis (Gahegan 2000; Andrienko et al. 2003; Guo et al. 2005; Laube et al. 2007; Kveladze et al. 2015). A GIS design which is capable of accommodating an integrated space-time system will provide a useful environment to visualize and manipulate trajectory data that are represented as space-time paths. When only a small number of space-time paths are involved in a study, it is quite easy to identify the stations formed among the paths through visualizing the paths in a space-time GIS environment. However, when a large number of paths are involved, the visualization scene becomes cluttered and it is impossible to discern any useful patterns. Methods are needed to restructure the data and provide simplified and intuitive visualization of the data to help researchers explore the stations. In this section, an aggregation approach which implements the station concept to effectively aggregate paths is proposed to help researchers explore stations existing in large tracking datasets of moving objects. An intuitive visualization of stations is also provided in the space-time GIS, which is discussed in Sect. 4, to help researchers comprehend the spatiotemporal characteristics of derived stations.

From an analytical perspective, a station is a location where a number of objects can bundle in space and time for certain events. A larger number of objects or a longer total stay time duration of the objects at a location usually indicates a higher significance level of the location as a station (Andrienko et al. 2003). There are many measures to define the significance level of a location as a station. In this study, the significance level of a station is defined by a magnitude measure which is the accumulated duration of all objects staying at the same location during a certain time period. This magnitude measure is used to evaluate the significance level of locations and detect potential station sites where moving objects form clusters. As discussed in Sect. 2, a space-time path is composed of a sequence of vertical and tilted line segments, which indicate an object's stays at specific locations and moves between locations respectively. Thus, the magnitude of a potential station can be derived by aggregating the vertical segments of multiple paths found at a specific location during a certain time period. Since the spatial and temporal extent of stations can vary significantly, choosing appropriate spatial and temporal resolution levels becomes essential to the exploration of stations. Several spatial and temporal aggregation systems are introduced in this study to provide more flexibility in exploring potential stations at various spatial and temporal structures and resolution levels.

As the first step to derive stations from path, the spatial extent for the aggregation process needs to be determined. Different tracking datasets may record the location information in different formats and at various spatial resolution levels. Also, researchers may have various familiarity levels on the movement patterns of the concerned moving objects. If researchers have developed good understanding of the movement patterns of the objects, they can develop a list of candidate sites for stations and investigate the clusters of trajectories at those sites. When researchers have little knowledge of the moving objects, they will have to rely on the dataset to explore the station sites. Therefore, different spatial aggregation methods may be needed under different circumstances. Three different spatial aggregation methods are proposed in this study to aggregate the recorded trajectories and explore where they cluster, including aggregating the paths based on (1) fixed spatial units, (2) spatial proximity defined by distance, and (3) spatial extent defined by kernel density estimation (KDE) analysis.

In some tracking datasets, a pre-defined fixed spatial unit system may be used in recording the movements of the observed objects. Such a fixed spatial unit system could be zip code tabulation areas, traffic analysis zones, counties, or a custom defined grid system. If it is meaningful to use a pre-defined fixed spatial unit to describe the distribution of the moving objects under investigation and represent the spatial extent of their stations, the fixed spatial units can be used to guide the aggregation of the paths. It is quite common that a pre-defined fixed spatial unit system may have a built-in hierarchical structure. For example, the administration boundary system contains several levels, including regions, states, and counties. The existing hierarchical structure of a fixed spatial unit system then can be used to support the exploration of the stations presented in a dataset at various spatial resolution levels.

In some cases, a pre-defined fixed spatial unit system may not be appropriate to guide the aggregation process. If prior knowledge of potential locations as stations exists, the distance-based spatial proximity method can be used to aggregate the paths. Under this approach, several places will be selected as the potential station sites first. Based on the existing knowledge of a potential site, a search radius can be determined to define the spatial extent for the potential station. A buffer zone will be calculated for each site and used to delineate the boundary of a potential station. All space-time paths that fall within the proximity defined by the buffer zone of a site will be aggregated to evaluate the significance of the site as a station. For instance, a half-mile circle centered at a city square can be used to define the spatial extent of the square as an activity station. Different distances may need to be tested in order to help researchers find the appropriate spatial extent to describe the stations in a given application.

The previous two methods are helpful when researchers already have some knowledge of the potential station sites. However, it is quite often that there is little knowledge of the spatial distribution pattern of the recorded moving objects. In this case, kernel density estimation (KDE) analysis can be used to identify the potential station sites and determine their spatial extents. In this approach, all locations visited by the moving objects are included in the analysis. A visit to a location from a moving object can be defined as the object stays at the location for a certain duration to complete an activity. Then the total stay duration at each location can be calculated by adding up the stay durations from all objects that have visited the location during the observation time period. For instance, if a place is visited twice by an object for 15 and 45 min respectively, and visited once by another object for 30 min, the total stay duration of this place from both objects will be 90 min. The total stay duration is then assigned as the weight of the location in KDE analysis. A density surface can be generated with a proper search radius for KDE analysis and "hotspots" can be identified with a chosen threshold of density level. Different from the previous two methods which use arbitrarily determined boundaries to delimit the location and spatial extent of station sites, this method allows the data to present itself for identifying the station sites. As a result, the derived stations may vary significantly in term of their spatial extent sizes.

After the spatial location of a station is determined, two temporal aggregation methods—fixed time interval method and moving time window method—can be applied to the dataset to investigate the variation of the magnitude of a station over time. The fixed time interval method divides the time span of a dataset into several time periods based on a user-specified time interval, such as a one-day time interval or a five-year time interval. The moving time window method, on the other hand, starts with a time window chosen by a user and creates the next time window by replacing the earliest year in the current time window with the year following the last year in the current time window. For example, a three-year moving time window will create time periods such as 2010–2012, 2011–2013, 2012–2014, etc. Once the time periods are defined (by either the fixed time interval or the moving time window method), the vertical segments of all objects' paths that fall within the spatial extent of a station site and a specific time period are aggregated to calculate the magnitude of the station for that particular time period. The variation of a station's significance over time then can be examined via the sequence of magnitudes calculated for the station site at each defined time period. As a result, the moving time window method usually creates smoother transitions between the adjacent time periods. Different sizes of time intervals and time windows can be tested before a decision is made on an appropriate temporal resolution level for stations in an application.

In the approaches discussed so far, a space-first-and-time-second strategy is implied for exploring the stations. Following this strategy, the spatial extents of stations are defined first (using fixed spatial units, proximity defined by distance, or spatial extent defined through KDE analysis) before the magnitude changes of these stations are examined. This strategy is based on an assumption that the spatial extent of a station remains unchanged through the observation time period. However, it is quite common that the spatial extent of a station changes over time. For example, the urbanized area of a city could expand over time through an urban sprawl process and the habitat area of a group of wild animals may migrate to different locations with seasonal changes. Therefore, it is also necessary to explore the spatial extent changes of stations while examining their magnitude variations. To achieve this goal, a time-first-and-space-second strategy can be implemented. Following this strategy, either the fixed time interval method or the moving window method is used to divide the tracking data first, and then a KDE analysis is applied to each subset of the tracking data in order to identify the spatial locations and extents of the hotspots (stations) for that particular time period. Different from the space-first-and-time-second strategy, this approach may produce hotspots with different locations and spatial extents for each time period. By assembling the hotspots from all time periods, it is now possible to examine the evolution of the identified stations in space and time.

Implementing the Station Concept in a Space-Time GIS Environment

This section introduces implementation of the proposed aggregation approaches and the time-geography station concept in a space-time GIS environment. A sample tracking dataset is used to demonstrate how such a GIS environment can reveal stations presented in the trajectories and support the visualization of the stations to help researchers comprehend the spatiotemporal characteristics of stations derived from the dataset. The 3D environment of ArcGIS, which is a product of the Environmental Systems Research Institute (ESRI), is adopted and adapted to simulate the space-time system of time geography. The third dimension (z) is used to represent the time dimension (t). The trajectory of a moving object then can be modeled as a 3D linear feature composed of a sequence of (x,y,t) triplets.

In this space-time GIS design, 3D cylinders are used to represent and visualize stations and a station is modeled as a sequence of cylinders for fixed spatial units or spatial extent defined by distance. All cylinders associated to a station will have their centers located at the same location, which can be either the exact location of a point station or the centroid location of an area station. The height of a cylinder indicates the duration of the defined time period used in the aggregation process, with the bottom surface of the cylinder located at the starting time of the period and the top surface at the ending time. The radius of a cylinder is used to represent the magnitude of the station in the specific time period, which is indicated by the position of the cylinder along the time dimension. A larger cylinder indicates more objects gathered at the site at the time, and the site is more likely to be an important site for the observed objects. The varying size of cylinders captures the evolution of a site's significance level as a host location for events associated to the objects. Such a space-time GIS representation of stations offers intuitive and convenient visualization to help researchers comprehend the dataset and explore the important locations associated with the observed moving objects.

The aggregation methods are applied to a sample individual-based travel and activity survey data and the derived stations are visualized in the space-time GIS environment for a proof-of-concept study. The sample travel and activity survey data is a subset of the travel tracker survey data collected for northeastern Illinois between January 2007 and February 2008. This subset dataset contains detailed travel inventory of 658 individuals (a total of 3510 recorded trips and activities) who participated in the one-day survey and had at least one trip recorded. Even though the sample data size is not very large, it is used to showcase how the analysis functions and visualization of stations work in the space-time GIS environment, and the functions and visualization can be readily applied to a large dataset. Each record in the survey dataset contains information such as a unique ID for each individual, location visited by the person, when the person arrived at and left the location. Due to privacy concerns, the location information has been aggregated to the census tract level when the data was released to the public. In the data preparation stage, all records belonging to the same individual in the sample dataset are extracted and sorted by time. The location and temporal information in these records is then used to construct a space-time path for the individual. In this process, all locations visited by an individual and his/her stays at those locations during the day are connected in their temporal order to form a 3D linear feature which can be stored in a new dataset in the space-time GIS. As shown in Fig. 2, a total of 658 space-time paths are generated to represent the trajectories of the surveyed individuals. The paths are then used to support spatiotemporal analyses in the space-time GIS for exploring stations presented in the trajectories.

In the first attempt to explore the dataset, each census tract is considered as a potential station as it is the spatial unit for reporting activity locations in the sample dataset, and its magnitude change is examined over the survey time period. The census tract boundary (a fixed spatial unit) and five-minute time interval (a fixed time interval) are used to aggregate the paths that represent the individuals travel activity patterns during the survey day. In this analysis process, a co-existence spatiotemporal relationship is examined among the constructed space-time paths at each census tract using the method developed by Yu (2006). The magnitude of each census tract at each time interval is calculated by accumulating a total stay duration of all paths staying at this location. A cylinder with a radius representing the level of magnitude is then generated and positioned at the centroid of the census tract and the correct time location in the space-time GIS.

Figure 3 shows the space-time GIS visualization of the stations derived from aggregating the paths. Only census tracts with a significant magnitude level are included in the figure. The size of the cylinder associated with a station varies as the station's significance level increases or decreases over time. From the visualization, one can tell that a few census tracts located in the downtown area of Chicago are

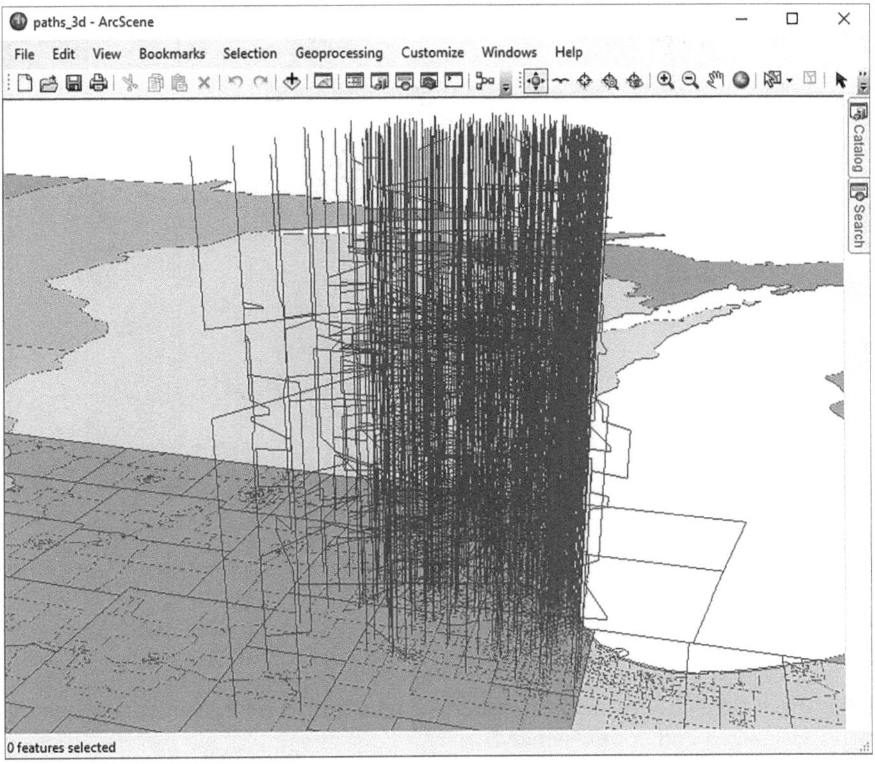

Fig. 2 Sample travel activity survey data represented as space-time paths

represented with a sequence of cylinders that show an increasing and then decreasing trend through the day. In the highlighted case located in the right of the figure, the magnitude level of this location (labeled as s1 in Fig. 3) starts to increase quickly around 8:40 am and maintains high through the day. Its magnitude level starts to decrease quickly at about 4:10 pm. This area is part of the central business district of Chicago, which has many jobs but fewer homes. Many individuals travel to this area for work during the day and leave for home in the evening. The magnitude level changes of this station correctly capture the characteristics of this location for work related activities. Many census tracts located in the suburbs are represented with a sequence of cylinders that have higher magnitude levels at both ends of the day. In the highlighted case (labeled as s2 in Fig. 3) located in the left of the figure, this census tract has high magnitude levels before 7:50 am and after 4:20 pm, and very low magnitude levels in between. As an area with many homes but fewer jobs, people leave this area for work during the day time and will not come back home until evening. The shape variation of the station depicts a typical place for home related activities. With restructured information and its visualization

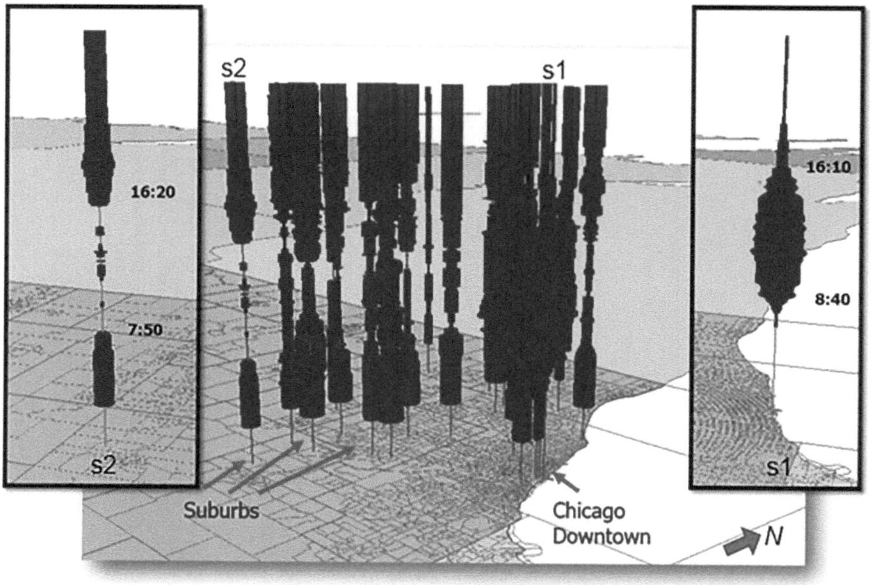

Fig. 3 Visualization of derived census tract-based activity stations in space-time GIS

in the space-time GIS, the activity patterns embedded in the tracking dataset can be easily visualized and comprehended.

The visualization of the aggregation results based on fixed spatial units offers an effective way to examine the activity patterns at the census tract level. However, the census tract boundary line may not be an appropriate choice to delineate the spatial extent of different types of activities, because the sizes of census tracts may vary significantly (much smaller in the downtown area and very large in the suburban area). In the second attempt, the spatial extent of major activity clusters in this area is determined by the activity location distribution presented in the dataset itself, instead of the fixed census tract boundaries. Therefore, the station boundary lines will not be limited to the census tracts. The second attempt will also examine the variations of the spatial extent of these stations over the survey time period. A 20-minute fixed time interval is used to divide the data into smaller subsets and KDE analysis is applied to each of the smaller subsets. A spatiotemporal dynamic segmentation method applied to space-time path (Yu 2006) is used to generate the sub-segments of the trajectories at the defined time interval. Each sub-segment is then converted to a set of 3D points at a finer temporal resolution (e.g., 5 min or 1 min) for KDE analysis. In order to run KDE analysis, a search radius needs to be determined for density calculation. Based on the results of several test runs, a search radius of 3.5 km is chosen as an appropriate radius for the analysis. This radius is about twice of the average size of census tracts in the surveyed area. After a density surface is generated, an equal interval classification method is used to classify the density values into several groups. A threshold value is then selected for identifying the "hotspot" locations (i.e., potential stations). Later, a cylinder is generated for each defined "hotspot" location, with the cylinder base shaped as the spatial extent of the "hotspot" and its height as the defined time interval. The cylinders are placed in their corresponding positions in the space-time GIS to represent the spatiotemporal characteristics of the identified stations.

Figure 4 shows the visualization of the stations derived from the KDE analysis approach. As shown in this figure, these stations are not composed of strict cylinders (whose intersections are circles), but a sequence of broadly defined cylinders (whose intersections can be irregular shapes). Each of these broadly defined cylinders is derived by extruding the polygon which delineates the spatial extent of the station at a specific time period along the time dimension according to the pre-defined time interval. Similar to the strict cylinders used in Fig. 3, the bottom surface of a broadly defined cylinder is located at the starting time of the period and the top surface at the ending time mark. However, different from the cylinders in Fig. 3 where the size of a cylinder indicates the magnitude of a station at a specific time period, the size of a broadly defined cylinder in Fig. 4 shows the spatial extent of a station. The varying sizes of the cylinders portray the location changes of a station over time.

The results shown in Fig. 4 indicate that different number of stations can be identified in the study area by choosing different levels of density threshold. The higher density threshold level is chosen, the fewer stations are identified (see Fig. 4a–c).

As shown in Fig. 4a, a station may not exist for the entire observation time period and its spatial extent may vary over time. In comparison to the census tract-based station results, the station (labeled as s1 in Fig. 4a) in the Chicago Downtown area now has a larger and changing spatial extent. It only shows up during the day time and is not recognized as a "hotspot" activity location in the early morning and late evening times. Again, this captures the place as a heavy work-related activity location. There are several stations (labeled as s2 and s3 in Fig. 4a) appear only in the early morning and late evening times. They are located in places that have a heavy presence of residential homes, where home-related activities are the major theme. There is one identified station (labeled as s4 in Fig. 4a) whose life time spans the whole day. This station is in an area with mixed land use types (residential and commercial). The combination of home-related activities in the early morning and late evening times and work-related activities during day time makes this place occupied with a significant level of clusters of people through the day. As the density level is calculated and compared through the entire survey area, home activity locations in the suburban areas, which usually have a more spread-out distribution pattern, are not captured in the KDE-based approach due to the very large magnitude level of the downtown area.

It is important to point out that these two approaches to implementing the station concept do not necessarily produce distinct results. Both the spatiotemporal cylinder and the KDE approaches yield a cluster of significant activity stations that

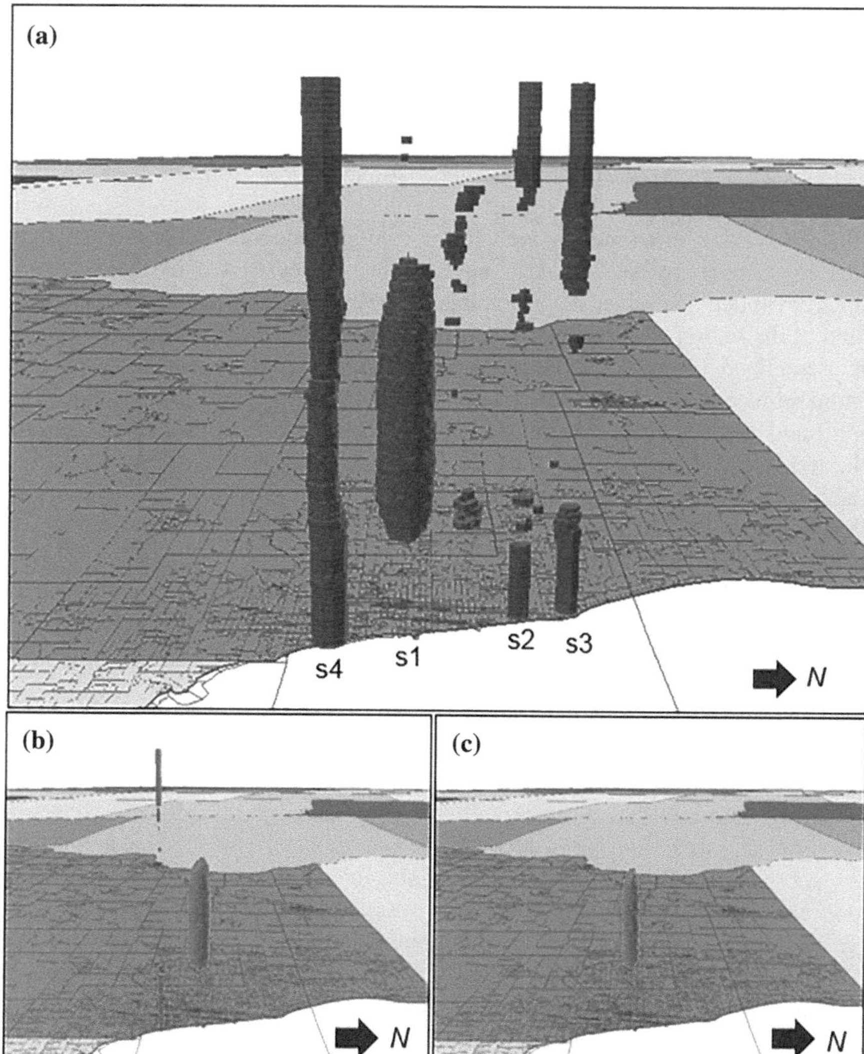

Fig. 4 Visualization of changing stations derived from KDE analysis in space-time GIS. Figure 4a shows stations defined at density level 2 and above; Fig. 4b shows stations defined at density level 3 and above; Fig. 4c shows stations defined at density level 4 and above

are centered in the Chicago Downtown area. Factors that may influence the choice of one approach over the other include whether or not the station sites are known and can be identified and whether the intended results are point-specific or area-based. Both approaches have demonstrated their usefulness to help researchers gain insight into patterns hidden in large individual tracking datasets.

Conclusions

Based on the station concept of time geography, this study proposes a system of aggregation methods in a space-time GIS environment to explore the important locations where the paths of many moving objects cluster in space and time. Several spatial and temporal aggregation methods are introduced to provide flexibility for researchers to manipulate the data and to identify stations with various spatial and temporal extents. Depending on whether researchers have some knowledge of the potential station sites, different representation approaches are proposed. When certain spatial locations have been identified as potential stations according to a priori knowledge, a sequence of cylinders centered at the centroid of an area spatial unit (e.g., a county) or at the exact location of a point unit (e.g., a city) are used to represent the stations. The varying sizes of the cylinders indicate the magnitude changes of the place as a station where the observed moving objects cluster. When little is known about the potential station sites of a group of observed moving objects, a sequence of broadly defined cylinders derived from extruding the polygons in the generated KDE surfaces are employed to visualize locations where space-time paths cluster. The proposed aggregation methods provide an effective approach to restructuring the trajectory data in large tracking datasets and exploring where and when the observed moving objects cluster. Representing stations as 3D objects, the space-time GIS design presents a useful and effective geovisualization environment to investigate the spatiotemporal characteristics of stations. With these capabilities, the proposed methods can benefit various research fields that utilize large tracking data sets for analysis.

At this moment, the proposed methods are designed to explore stations that are defined by the clusters of the vertical segments of space-time paths. In other words, the approach can only capture the bundles of space-time paths at fixed locations. As researchers have acknowledged, space-time paths may bundle either at fixed locations (e.g., buildings) or during movements (e.g., car-pooling). While the first scenarios are known as stationary bundles, the latter ones are referred as mobile bundles (Miller 2004). In order to explore mobile bundles, the tilted segments of space-time paths need to be included in the analysis. As the tilted segments of space-time paths may have numerous choices of directions in the space-time system and it becomes more complex when defining proximity among a titled space-time path segment in the space and time system, it presents an even more challenging research problem. However, being able to identify the mobile bundles among a large number of space-time paths is very important in some studies such as pinpointing where and when the vehicles on a road start to converge and form traffic congestion. For future development directions, the proposed methods need to be expanded so that they can be used to investigate both stationary and mobile bundles among space-time paths and provide enhanced analysis power to explore the spatiotemporal clusters of trajectories in large tracking datasets.

References

Andrienko, G., Andrienko, N., & Gritis, V. (2003). Interactive maps for visual exploration of grid and vector geodata. *Photogrammetry and Remote Sensing, 9*(2), 380–389.

Andrienko, G., Andrienko, N., & Wrobel, S. (2007). Visual analytics tools for analysis of movement data. *ACM SIGKDD Explorations, 9*(2), 38–46.

Brinkhoff, T. (2002). A framework for generating network-based moving objects. *GeoInformatica, 6*(2), 153–180.

Buliung, R. N., & Kanaroglou, P. S. (2006). A GIS toolkit for exploring geographies of household activity/travel behavior. *Journal of Transport Geography, 14,* 35–51.

Dodge, S., Weibel, R., & Forootan, E. (2009). Revealing the physics of movement: Comparing the similarity of movement characteristics of different types of moving objects. *Computers, Environment and Urban Systems, 33*(6), 419–434.

Dykes, J. A., & Mountain, D. M. (2003). Seeking structure in records of spatiotemporal behavior, visualization issues, efforts and applications. *Computational Statistics & Data Analysis, 43,* 581–603.

Erwig, M., Güting, R., Schneider, M., & Vazirgiannis, M. (1999). Spatio-temporal data types, an approach to modeling and querying moving objects in databases. *GeoInformatica, 3*(3), 269–296.

Gahegan, M. (2000). The case for inductive and visual techniques in the analysis of spatial data. *Journal of Geographical Systems, 2,* 77–83.

Golledge, R., & Stimson, R. (1997). *Spatial Behavior: A Geographic Perspective.* New York: The Guilford Press.

Guo, D., Gahegan, M., MacEachren, A. M., & Zhou, B. (2005). Multivariate analysis and geovisualization with an integrated geographic knowledge discovery approach. *Cartography and Geographic Information Science, 32,* 113–132.

Güting, R., Böhlen, M., Erwig, M., Jensen, C., Lorentzos, N., Schneider, M., et al. (2000). A foundation for representing and querying moving objects. *ACM Transactions on Database Systems, 25*(1), 1–42.

Hägerstrand, T. (1970). What about people in regional science? *Papers of the Regional Science Association, 24,* 7–21.

Kulldoff, M. (2001). Prospective time periodic geographical disease surveillance using a scan statistic. *Journal of the Royal Statistical Society Series A, 164,* 61–72.

Kveladze, I., Kraak, M. J., & Van Elzakker, C. P. (2015). The space-time cube as part of a GeoVisual analytics environment to support the understanding of movement data. *International Journal of Geographical Information Science, 29*(11), 2001–2016.

Kwan, M.-P. (2000a). Human extensibility and individual hybrid-accessibility in space-time, a multi-scale representation using GIS. In D. Janelle & D. Hodge (Eds.), *Information, place, and cyberspace, issues in accessibility* (pp. 241–256). Berlin, Germany: Springer-Verlag.

Kwan, M.-P. (2000b). Interactive geovisualization of activity-travel patterns using three dimensional geographical information systems, A methodological exploration with a large data set. *Transportation Research C, 8,* 185–203.

Kwan, M.-P., & Hong, X. (1998). Network-based constraints-oriented choice set formation using GIS. *Geographical Systems, 5,* 139–162.

Laube, P., Dennis, T., Forer, P., & Walker, M. (2007). Movement beyond the snapshot–dynamic analysis of geospatial lifelines. *Computers, Environment and Urban Systems, 31*(5), 481–501.

Laube, P., & Purves, R. S. (2006). An approach to evaluating motion pattern detection techniques in spatio-temporal data. *Computers, Environment and Urban Systems, 30,* 347–374.

Long, J. A., & Nelson, T. A. (2013). A review of quantitative methods for movement data. *International Journal of Geographical Information Science, 27*(2), 292–318.

Miller, H. (1991). Modeling accessibility using space-time prism concepts within geographical information systems. *International Journal of Geographical Information Systems, 5,* 287–301.

Miller, H. (2004). Activities in space and time. In D. Hensher, K. Button, K. Haynes, & P. Stopher (Eds.), *Handbook of transport 5, transport geography and spatial systems* (pp. 647–660). London, UK: Elsevier Science.

Miller, H. (2005). A measurement theory for time geography. *Geographical Analysis, 37*(1), 17–45.

Neutens, T., Van de Weghe, N., Witlox, F., & De Maeyer, P. (2008). A three-dimensional network-based space-time prism. *Journal of Geographical Systems, 10*(1), 89–107.

Onozuka, D., & Hagihara, A. (2007). Geographic prediction of tuberculosis clusters in Fukuoka, Japan, using the space-time scan statistic. *BMC Infectious Diseases, 7,* 26–34.

Parkes, D., & Thrift, N. (1980). *Times, spaces, and places: A chronogeographic perspective*. New York: Wiley.

Porkaew, K., Lazaridis, I., & Mehrotra, S. (2001). Querying mobile objects in spatio-temporal databases. *SSTD, 2001,* 59–78.

Postlethwaite, C. M., Brown, P., & Dennis, T. E. (2013). A new multi-scale measure for analysing animal movement data. *Journal of Theoretical Biology, 317,* 175–185.

Pred, A. (1977). The choreography of existence: Comments on Hägerstrand's time-geography and its usefulness. *Economic Geography, 53*(2), 207–221.

Purves, R. S., Laube, P., Buchin, M., & Speckmann, B. (2014). Moving beyond the point: An agenda for research in movement analysis with real data. *Computers, Environment and Urban Systems, 47,* 1–4.

Rinner, C. (2004). Three-dimensional visualization of activity-travel patterns. In M. Raubal, A. Sliwinski & K. Kuhn (Eds.), Geoinformation und Mobilität [Geoinformation and Mobility], Proc. of the Münster GI Days, 1–2 July 2004, Münster, Germany, IfGIprints series No. 22. Verlag Natur and Wissenschaft, Solingen, Germany, pp. 231–237. http://www.ryerson.ca/~crinner/pubs/rinner-3d-vis_full.pdf. Accessed on Jan 8, 2008.

Shaw, S.-L., & Yu, H. (2009). A GIS-based time-geographic approach of studying individual activities and interactions in a hybrid physical-virtual space. *Journal of Transport Geography, 17*(2), 141–149.

Shaw, S.-L., Yu, H., & Bombom, L. S. (2008). A space-time GIS approach to exploring large individual-based spatiotemporal datasets. *Transactions in GIS, 12*(4), 425–441.

Vazirgiannis, M., & Wolfson, O. (2001). A spatiotemporal model and language for moving objects on road networks. *SSTD, 2001,* 20–35.

Wolfson, O., Xu, B., Chamberlain, S., & Jiang, L. (1998). Moving objects databases, issues and solutions. *Proceedings of SSDB Conference, 1998,* 111–122.

Yu, H. (2006). Spatio-temporal GIS design for exploring interactions of human activities. *Cartography and Geographic Information Science, 33*(1), 3–19.

Yu, H., & Shaw, S.-L. (2008). Exploring potential human activities in physical and virtual spaces, a spatio-temporal GIS approach. *International Journal of Geographical Information Science, 22*(4), 409–430.

Yuan, M., Mark, D., Egenhofer, M., & Peuquet, D. (2004). Extensions to geographic representations. In R. McMaster & E. Usery (Eds.), *A research agenda for geographic information science* (pp. 129–156). Boca Raton, FL: CRC Press.

An Extended Community Detection Algorithm to Compare Human Mobility Flow Based on Urban Polycentric Cluster Boundaries: A Case Study of Shenzhen City

Zhixiang Fang, Lihan Liu, Shih-Lung Shaw and Ling Yin

Introduction

Batty's (2013) new science of cities strengthens the focus on flow and network, which is "*the full story of how cities grow and evolve into different forms and functions.*" Urban flow and networks are relationships between people and places, which are affected by urban spatial structure, including forms and functions. Many studies have shown the advantages of polycentric urban structure (Handy 1996; Schwanen et al. 2001; Parr 2004) for improving travel efficiency, reducing traffic

Z. Fang · L. Liu · S.-L. Shaw
State Key Laboratory of Information Engineering in Surveying, Mapping and Remote Sensing, Wuhan University, Wuhan 430079 People's Republic of China
e-mail: zxfang@whu.edu.cn

S.-L. Shaw
e-mail: sshaw@utk.edu

Z. Fang · S.-L. Shaw
Collaborative Innovation Center of Geospatial Technology, Wuhan 430079, People's Republic of China

L. Liu (✉)
Chengdu Institute of Planning and Design, Chendu 610041 People's Republic of China
e-mail: liulihan@whu.edu.cn

S.-L. Shaw
Department of Geography, University of Tennessee, Knoxville, TN 37996, USA

L. Yin
Shenzhen Institutes of Advanced Technology, Chinese Academy of Sciences, Shenzhen 518005, People's Republic of China
e-mail: yinling@siat.ac.cn

C. Zhou et al. (eds.), *Spatial Data Handling in Big Data Era*, Advances in Geographic Information Science,
DOI 10.1007/978-981-10-4424-3_8

flows, and saving energy. Polycentric urban spatial structure is currently viewed as an efficient urban structure to attract and service large populations in developing countries.

Before designing a sustainable polycentric spatial structure, urban planners or agencies need to examine the differences between human mobility flow communities and the polycentric cluster centers, e.g. the community boundaries and cluster center service areas. However, this is significant challenge because there is lack of detailed human mobility flow data covering a large proportion of the urban residents. To address this challenge, this paper uses mobile phone location data, which is developing as a data source of great importance in urban planning field.

The European spatial development perspective launched in 1999 is a popular policy for European member states (Krätke 2001). Many researchers have studied polycentric urban forms for commuting patterns and behaviors, employment, housing, etc. Kloosterman and Musterd (2001), and Kloosterman and Lambregts (2001), showed that polycentricity can refer to intra-urban clustering patterns for population and economic activity. Dieleman et al. (2002) investigated the urban form and travel behavior from micro-level household attributes and residential context. Meijers and Romein (2003) argued that potential planning for polycentric urban regions requires active development of regional organizing capacity, which should be influenced by spatial, functional, political, institutional, and cultural factors. Meijers (2005) found that polycentric urban forms can perform better than the sum of their parts via cooperative and complementary relationships. Yue et al. (2010) investigated polycentric urban development through analyzing the directions of urban expansion, urban-rural gradients, and growth types. Modarres (2011) investigated the commuting patterns of polycentric cities in Southern California, and suggested that "*advocacy for any particular urban form may be premature and less than efficient if we do not take into account the reality of commuting patterns as they relate to our fragmented and decentered metropolitan areas*" (p. 1193). Grunfelder and Nielsen (2012) investigated the relationship between urban form and commuting behavior in a polycentric urban region, and found that "*the distance to the closest urban center is an important factor affecting commuting. In the aspect of employment, polycentric urban employment patterns may provide a better explanation of commuting patterns*". McDonald and McMillen (1990) investigated employment subcenters and land values in a polycentric urban area. Redfearn (2007) introduced a nonparametric method of identifying subcenters of employment in polycentric urban areas. Han (2005) explored the spatial clustering of property values in polycentric urban development by global and local spatial auto-correlation. Wen and Tao (2015) found that urban planning policy and housing market forces drove polycentric urban development in Hangzhou city. These studies also demonstrated the advantages of the polycentric urban form. However, the usefulness and sustainability of the polycentric urban form must be tested and validated (Meijers, 2008), which has been addressed by a number of researchers. Vasanen (2013) showed that the degree of functional polycentricity varies considerably across different spatial scales. Brezzi and Veneri (2015) provided measures of polycentricity and explored the economic implications of different spatial structures. Roth et al. (2011) used a large scale, real time 'Oyster' card database of individual movements in the London subway to examine the structure and organization of the city. However, difference patterns between human mobility flow communities and urban cluster structure need to be further examined to improve planned polycentric urban structures.

Community is an important concept in geography, sociology, biology, and computer science. Identifying community boundaries assists with understanding the divisions of human mobility and class, culture, racial or ethnic status, etc. Several methods have been proposed to detect communities in graphs or networks. Lancichinetti and Fortunato (2009) compared community detection algorithms. Fortunato (2010) reviewed the community detection methods from graphs, including traditional methods (i.e., graph partitioning and clustering), modularity based divisive algorithms, spectral and dynamic algorithms, etc. Newman-Girvan modularity has become an essential element in many community detection (Li et al. 2008) or cluster methods, which is an often used greedy approach for clustering complex networks, such as social or human flow networks, etc. Duch and Arenas (2005) used external optimization to detect communities in complex networks. Barber (2007) defined a bipartite modularity for community detection, which was used to identify the modular structure of bipartite networks. Gog et al. (2007) proposed an evolutionary technique for community detection in complex networks on the basis of information sharing between population individuals. Gong et al. (2012) used a multi-objective optimization algorithm to detect community by simultaneously maximizing the density of internal degrees and minimizing the density of external degrees. Yang et al. (2013) proposed the communities from edge structure and node attributes (CENA) algorithm to detect overlapping communities in networks with node attributes, which improved the accuracy and robustness for the case of network structures. Zhou et al. (2013) proposed a partition method for community structure in complex networks based on edge density. Niu et al. (2013) proposed a complex network community detection algorithm based on core nodes. In application, Nan (2014) introduced a prediction for hot regions based on complex networks and community detection. Leung et al. (2009) investigated a real time community detection problem for large scale online social networks by incorporating different heuristics. Padopoulos et al. (2012) framed the problem of community detection in social media networks by acknowledging the unprecedented scale, complexity, and dynamic nature of the network, and provided a compact classification of existing algorithms. Shi et al. (2012) formulated a multi-objective framework for community detection in social network and proposed a multi-objective evolutionary algorithm for finding efficient solutions under the framework. Yin (2014) investigated local interested community detection in large scale social networks using the relationships of users' friends and microblogging users' interest information. Wang and Cheng (2015) investigated the dynamic community in online social networks for detecting abnormal swarm events. Few studies have addressed community detection algorithms with fixed sources. This chapter proposes an extended community detection algorithm of human mobility flow with the constraint that fixed sources are equal to the number of urban clusters. Thus, the planned urban structure can be compared with *corresponding* communities.

According to the above literature reviews, few studies have examined the differences between human mobility flow communities and planned urban cluster areas or centers. This paper addresses at this issue. Section "Proposed Human Flow Based Community Detection Algorithm" introduces a human flow based community detection algorithm with the constraint of fixed sources. Section "Case Study Area and Data" introduces the case study area and data, with results, analysis and discussion presented in Section "Results and Discussion". Section "Conclusion" presents the conclusions of the study.

Proposed Human Flow Based Community Detection Algorithm

To compare the difference of human flows on urban cluster centers, we need an algorithm capable of generating communities guided by the planned urban cluster centers. Due to the difficulty of State of the art community detection algorithms struggle to control community detection process with constraints of initializing source nodes. Therefore, we introduce a human flow based community detection (HFCD) algorithm with already partitioned source nodes, based on human mobility flow derived from mobile phone data, that can generate communities integrating the initialized source partitions. This algorithm is based on the hierarchical agglomeration algorithm of CNM (Clauset et al. 2004) designed for detecting community structure, which is faster than many state of the art competing algorithms.

Let $G=(V,E)$ represents a mobile communication base station network, where V is a set of base stations, and E is a set of links between them. Mobile phone location data records user trajectories as a series of mobile communication base stations with time stamps, where $Trj(i) = \{v_1,v_2,v_3,\ldots,v_{n-2},v_{n-1},v_n\}$ is the ith user's trajectory, $v_1 = (x_1,y_1,t_1)$, x_1 and y_1 are the longitude and latitude of the mobile communication base station, and t_1 is the time the user presents to the base station.

Step 1 Build G using all trajectories in the mobile phone location dataset. For each trajectory, $Trj(i)$, link each neighboring base station pair within this trajectory and update the total frequencies between these stations, i.e., pairs $\{v_1,v_2\},\{v_2, v_3\}, \ldots, \{v_{n-2},v_{n-1}\},\{v_{n-1}, v_n\}$. Thus, each base station pair has a movement frequency, which represents the human flow between them.

Step 2 Select mobile communication base stations in G as source nodes in the community. Divide the base stations in G into subsets using the official urban center boundaries, and then find the base station with the highest frequency for each divided subset. This initializes source nodes for further community detection, which differs from previous community detection approaches (Clauset et al. 2004; Lancichinetti and Fortunato 2009;

Fortunato 2010; Padopoulos et al. 2012; Zhou et al. 2013). Each source node is assigned to an initialized community, and the other nodes are assigned to a non-source community. Thus, each base station is assigned to a source or non-source community, C for the set of source communities, S. Figure 1 shows the source nodes in the community detection algorithm.

Step 3 Build a sparse matrix, ΔQ, for all mobile communication base stations, that represents the increment of modularity after two base stations are merged. Let

$$a_i = \frac{k_i}{2m},\ m = \sum_{i \in G} a_i, \tag{1}$$

then

$$\Delta Q_{ij} = \begin{cases} \frac{1}{2m} - \frac{k_i \times k_j}{(2m)^2} & \alpha_{ij} \neq 0 \\ 0 & \alpha_{ij} = 0 \end{cases}, \tag{2}$$

where v_i and v_j are nodes i and j, a_i is the weight of community i, k_i and k_j are the node degrees of nodes i and j, and m is the total weights of all edges in G. $\alpha_{ij} = 1$ when community i links with j, and $\alpha_{ij} = 0$ otherwise

Step 4 Find the maximal ΔQ_{ij}, while the i and j couldn't be both source communities. Merge community i and j, then update ΔQ and a_i. We need to update the column j and row j of ΔQ, then remove the column i and row i of ΔQ .There are three cases in this updating process:

Case 1: $\Delta Q'_{jk} = \Delta Q_{ik} + \Delta Q_{jk}$ when community k links communities i and j simultaneously.

Case 2: $\Delta Q'_{jk} = \Delta Q_{ik} - 2a_j a_k$ when community k links community i, but not j.

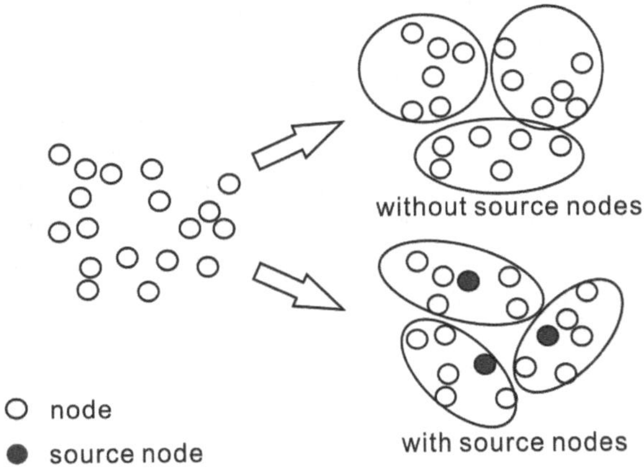

Fig. 1 Source nodes in the community detection algorithm

Case 3: $\Delta Q'_{jk} = \Delta Q_{jk} - 2a_i a_k$ when found *k* links community *j*, but not *i*.

After this merging, the weight of node *j* is updated as $a'_j = a_j + a_i$, $a_i = 0$, then the new community named *i*.

Step 5 Repeat step 4 until there is no $\Delta Q_{ij} \leq 0$, i.e., all nodes are assigned to detected communities. The source communities enlarge after merging some non-source nodes.

The proposed algorithm extends the CNM algorithm by including the source partitions to guide the community detection algorithm; introducing a separate node merging strategy between source and non-source nodes, but forbidding merging of communities containing source partitions; and using a sparse matrix data structure to maintain the community modularity attributes, which support the modularity update for each step. Therefore, the proposed algorithm can generate communities according to maximal modularity, which can be different from the source partitions. The maximal modularity criteria means the algorithm will still find communities that don't include source partitions.

The computational complexity of the proposed algorithm is $O(m(\log_2 n)^2)$, which is same as the CNM algorithm.

Case Study Area and Data

This study used Shenzhen city as the study area. Shenzhen is a major city and financial center in southern China, located immediately north of the Hong Kong Special Administrative Region, and was the first special economic zone (SEZ). In 2015, the GDP of Shenzhen was approximately $USD 270 billion. The SEZ included Luohu, Futain, Nanshan, and Yantian districts (Fig. 1) until 1 July 2010, then all Shenzhen city districts were included. Shenzhen has many high-tech companies and two main industrial parks, Shenzhen Hi-Tech Industrial Park and Shenzhen Software Park. Fig. 2 shows the administrative divisions of Shenzhen City, and Table 1 lists some of their demographic details, including area, population, subdistricts, and residential communities.

Shenzhen city government released its comprehensive plan (2010–2020) in September, 2010. The spatial development strategy included two development axes, three development belts, and polycenters and clusters or groups, as shown in Fig. 3. The orange arrows in the eastern and western areas are the development belts, and the blue arrows in the northern and southern areas are development axes. Futian-Luohu and Qianhai centers are important main centers with five subcenters and eight cluster centers planned to provide aggregated urban functions for residents.

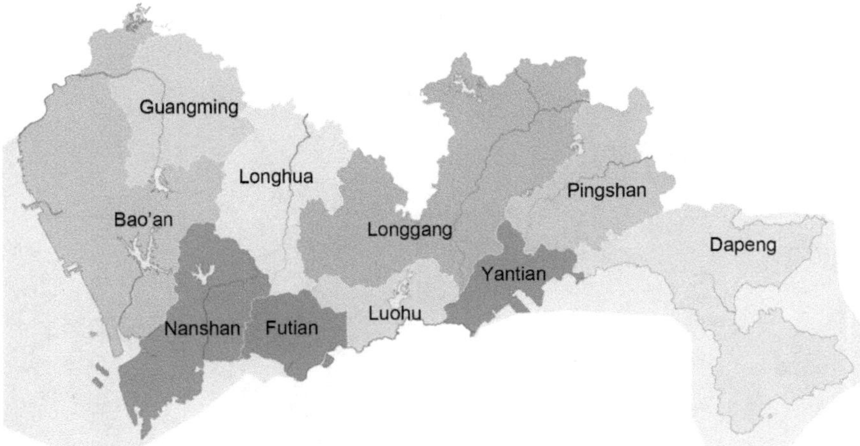

Fig. 2 Administrative divisions of Shenzhen City. *Source* https://en.wikipedia.org/wiki/Shenzhen/

Table 1 Administrative divisions of Shenzhen City

District	Area (km^2)	Population (2010)	Subdistricts	Residential communities
Luohu	78.75	923,421	10	115
Futian	78.65	1,317,511	10	114
Nanshan	185.49	1,088,345	8	105
Bao'an	398.38	2,638,917	6	266
Longgang	387.82	1,672,720	8	170
Yantian	74.63	209,360	4	22
Guangming	155.44	480,907	2	28
Pingshan	167.00	300,800	2	30
Longhua	175.58	1,379,460	6	100
Dapeng	295.05	126,560	3	25

Source https://en.wikipedia.org/wiki/Shenzhen/

A mobile phone location dataset was used to derive human flow, containing 1,627,265 anonymous users, and 43,414,969 records from 2841 mobile communication base stations. Each record includes user ID, date, time, and x and y location coordinates. Table 2 shows the seven land use types, and mobile communication base station totals for each land use. Residential (28.72%), transportation (24.29%), and industrial (22.74%) were the three major land uses including high percentages of mobile communication base stations for this dataset, as shown in Fig. 4. Approximately 38% of users called only once in the data set, and so could not be used to recover trajectories.

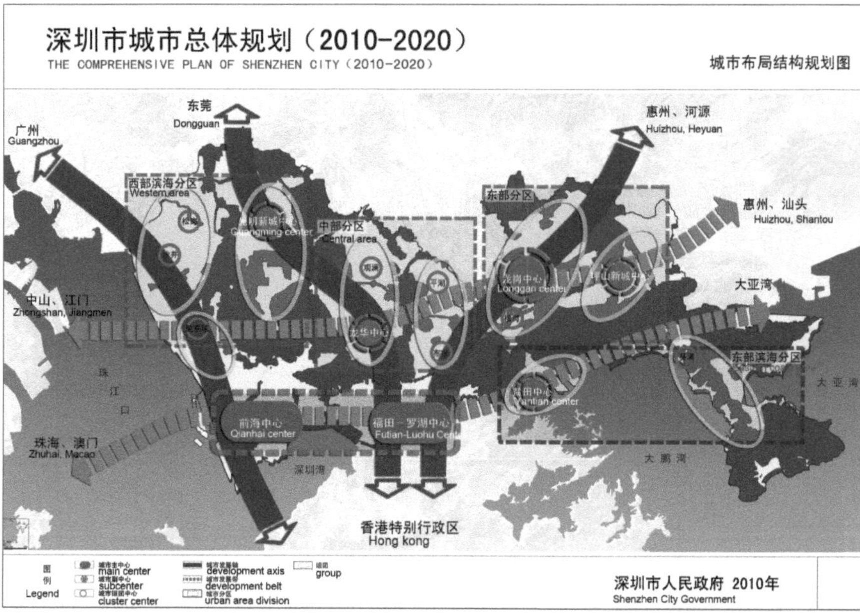

Fig. 3 Comprehensive plan for Shenzhen City (2010–2020). *Source* http://www.szfdc.gov.cn/szupb/

Table 2 Land use and mobile communication base station distribution in Shenzhen

Land use	Area (km^2)	Percentage (%)	Communication base number	Percentage (%)
Residency	19,406	9.82	816	28.72
Commerce	3024	1.63	173	6.09
Public service	9175	4.64	225	7.92
Transportation	22,084	11.18	690	24.29
Industry	29,862	15.12	646	22.74
Agriculture	92,714	46.93	155	5.46
Unused land	11,939	6.04	54	1.90

Results and Discussion

Figure 5 shows the selected source node set used for the proposed HFCD algorithm, shown as red areas. The number of initialized groups of source nodes was the same as the number of planned cluster centers from the Comprehensive plan. Twenty-three communities were by the proposed HFCD algorithm, for the highest modularity = 0.7198. These included seven new communities, which did not include any initialized source nodes. The new communities were identified by their

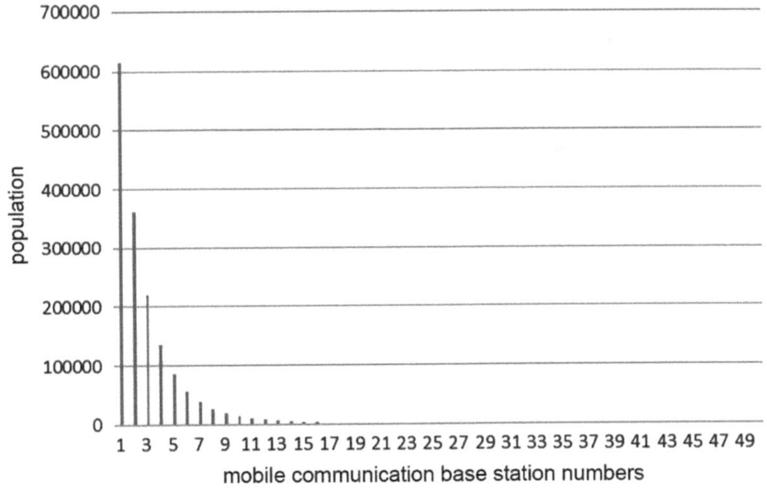

Fig. 4 Usage and mobile communication base station distributions

relatively high modularity from their surrounding areas, i.e., their nearby planned cluster centers do not act as key centers for residents.

Figure 6 shows the communities detected by the CNM algorithm; and Fig. 7 shows the planned center boundaries from Fig. 2. Table 3 compares the proposed and CNM algorithms. The minimal number of nodes was 21 and 25, and the maximal number was 338 and 277, respectively. The algorithms' modularities are similar. The proposed algorithm produces less difference of mumble of detected communities than the CNM algorithm. Thus, the proposed algorithm has better cluster balancing performance for these centers.

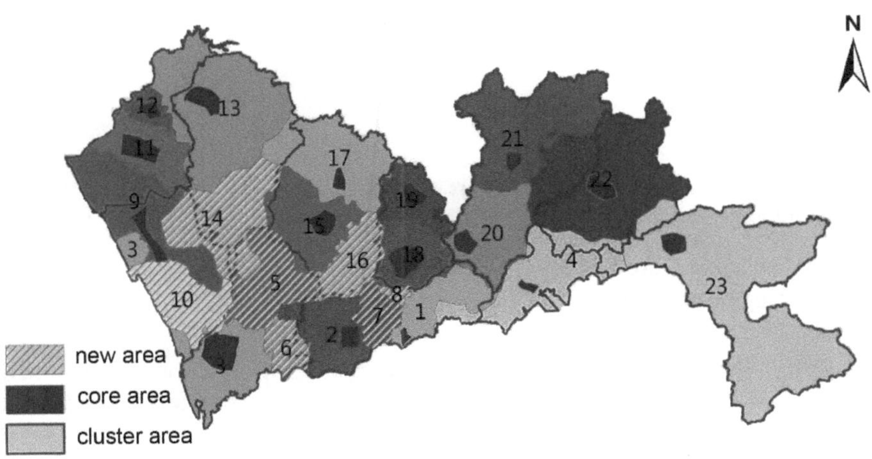

Fig. 5 Communities detected by the proposed HFCD algorithm

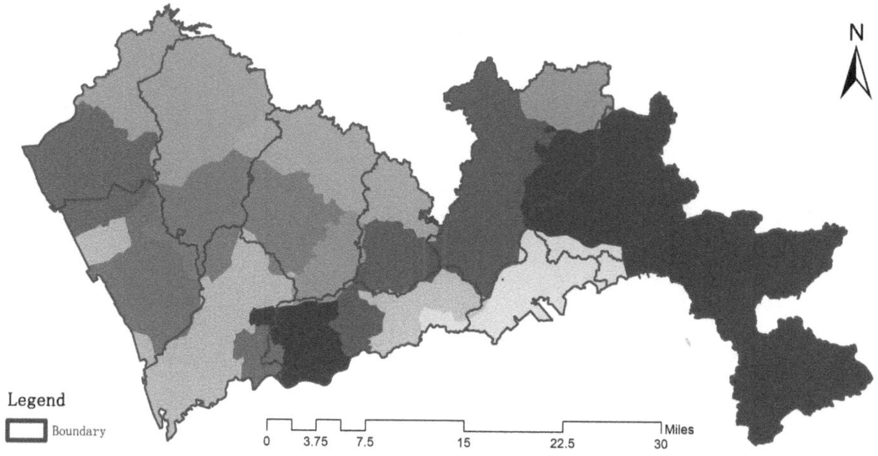

Fig. 6 Communities detected by the CNM algorithm

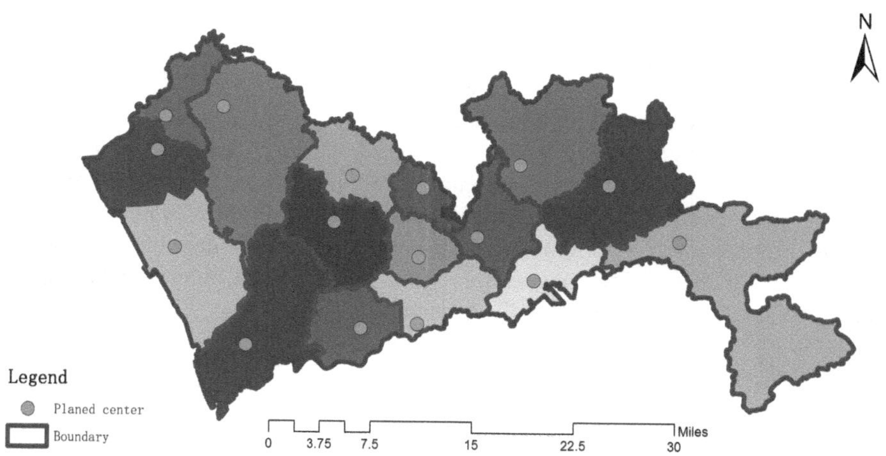

Fig. 7 Planned centers and boundaries

Table 4 compares the planned centers and detected communities. Some detected communities were merged to provide a valid comparison with the planned centers.

(1) Almost half of the detected communities have similar areas to their corresponding planned centers (difference <10 km^2) i.e., areas 3, 5, 6, 15, 16, 18–20, and 23 in Fig. 4. The similar size implies that human mobility flows in these areas follow the planned center boundaries. For example, area 3 is a new special economy center (Qianhai) within the Shenzhen urban area, areas 5 and 6 are new communities found by the proposed algorithm in the same general regions as area 3 (see Fig. 6), which implies that human mobility flows in the

Table 3 Nodes and communities detected using the proposed and CNM algorithms

Item	CNM algorithm	Proposed algorithm
Total nodes	2841	2841
Source nodes	—	426
Non-source nodes	2841	2415
Source community	—	16
Divided community	18	23
Minimal node number in community	21	25
Maximal node number in community	338	277
Modularity	0.7198	0.7155

Table 4 Area differences between planned centers and detected communities

Community	Planned area (km^2)	Detected community area (km^2)	Area difference (km^2)
1	78.5	54.9	23.6
2,7	78.5	88.9	−10.4
3,5,6	178.6	172.0	6.6
4	74.4	102.3	−27.9
8,9,10	161.5	177.9	−16.4
11	97.6	52.5	45.1
12	63.9	18.9	45
13,14	220.3	281.1	−60.8
15,16	114.5	122.8	−8.3
17	89	102.7	−13.7
18	55.4	53.2	2.2
19	40.9	44.6	−3.7
20	82.9	80.5	2.4
21	178.5	139.4	39.1
22	166.5	185.3	−18.8
23	294.5	298.4	−3.9

planned area do not follow the planned single center boundaries, but they are divided into three close parts under the construction periods. Areas 15, 16, 18–20, and 23 have very small area difference (<10 km^2). These are rural areas of Shenzhen city, and they show very little variance between the serviced and planned area, which implies the planned centers in these areas are successfully attracting human motilities.

(2) Several detected communities have large area difference from their corresponding planned centers (difference >35 km^2), i.e., areas 11–14, and 21 in Fig. 4. These areas are the most rural areas in the region, and are also the newly developing zones, e.g. area 13 is a new industrial area that includes three of the nine economic pillars in Shenzhen city, area 14 is a new high-tech development

zone including leisure areas near to a forest park, area 21 also includes a large forest, which attracts tourists and leisure activities. Human mobility flows in these areas are quite different from the planned centers, which implies that these rural centers have not become powerful service centers yet. The functions and facilities in these areas need to be carefully planned and efficiently implemented to support human motilities.

(3) Areas 3, 4, 9, 10, 17, and 22 have intermediate differences from their corresponding planned centers. They present three distinct patterns.

Areas 4 and 17 showed expanded area into the adjacent regions. This indicates that the planned center has less attraction than other centers, and suggests the planned center needs to strengthen its service functions.

Area 22 shows both reduced area and covering another center's area. This indicates that the planned centers do not have sufficient service ability to achieve their planned objectives, and these areas need to consider adjusting or improving their planned scheme to better service their own regions.

Areas 3, 9, 10 boundaries show little correspondence with the planned boundaries. These areas cover two planned centers, but three communities were detected, which implies the planned centers are mismatched with their mobility communities, and planning for these areas should be adjusted in future revisions.

(4) Areas 5–7, 10 and 14 do not correspond with planned centers. They are within the service areas of planned centers, but have relatively high modularity of human mobility flows. This implies that these areas have become a subcenter, and the relationship between these areas and their adjacent areas are needed to be reconsidered. Alternative strategies could be considered, such as subdividing these areas into centers, or strengthening the connections between these areas to better correspond to the planned center.

Conclusion

Urban human mobility has been intensively researched, but few studies have investigated communities of human mobility flows to estimate the implementation of planned polycentric urban structures, which is a challenge for urban planning. This paper proposes an extended community detection approach to compare detected community boundaries with planned polycentric cluster areas, which could be easily-implemented by urban planning agencies to support human mobility estimates. The case study in Shenzhen city showed that the proposed algorithm was effective at estimating the clustering of human mobility under the polycentric urban structure. Comparing detected human mobility community boundaries and planned polycentric cluster areas would help urban planning agencies to find areas that need to be improved in future planning stages.

Acknowledgements This study was jointly supported by National Nature Science Foundation of China (Grants #41231171, #41371420, #41371377, #41301511), the innovative research funding of Wuhan University (2042015KF0167), Arts and Sciences Excellence Professorship and Alvin and Sally Beaman Professorship at the University of Tennessee.

References

Barber, M. (2007). Modularity and community detection in bipartite network. *Physical Review E, 76,* 066102.

Batty, M. (2013). *The new science of cities.* Cambridge, Massachusetts: The MIT Press.

Brezzi, M., & Veneri, P. (2015). Assessing polycentric urban systems in the OECD: Country. *Regional and Metropolitan Perspectives, European Planning Studies, 23*(6), 1128–1145.

Clauset, A., Newman, M., & Moore, C. (2004). Finding community structure in very large networks. *Physical Review E, 70*(6), 66111.

Dieleman, F. M., Dijst, M., & Burghouwt, G. (2002). Urban form and travel behaviour: Micro-level household attributes and residential context. *Urban studies, 39*(3), 507–527.

Duch, J., & Arenas, A. (2005). Community detection in complex networks using external optimization. *Physical Review E, 72,* 027104.

Fortunato, S. (2010). Community detection in graphs. *Physics Reports, 486,* 75–174.

Gog, A., Dumitrescu, D., & Hirsbrunner, B. (2007). Community detection in complex networks using collaborative evolutionary algorithms. In F. A. e Costa, L. M. Rocha, E. H. Costa & I. A. Coutinho (Eds.), *Advances in artificial life, 9th European conference, ECAL 2007, Lisbon, Portugal* (pp. 886–894), September 10–14, 2007. Proceedings, Berlin: Springer-Verlag.

Gong, M., Ma, L., Zhang, Q., & Jiao, L. (2012). Community detection in networks by using multiobjective evolutionary algorithm with decomposition. *Physica A, 391,* 4050–4060.

Grunfelder, J., & Nielsen, T. S. (2012). Commuting behaviour and urban form: a longitudinal study of a polycentric urban region in Denmark. *Geografisk Tidsskrift-Danish Journal of Geography, 112*(1), 2–14.

Han, S. S. (2005). Polycentric urban development and spatial clustering of condominium property values: Singapore in the 1990s. *Environment Planning A, 37*(3), 463–481.

Handy, S. (1996). Methodologies for exploring the link between urban form and travel behavior. *Transportation Research Part D: Transport and Environment, 1*(2), 151–165.

Krätke, S. (2001). Strengthening the polycentric urban system in Europe: Conclusions from the ESDP. *European Planning Studies, 9*(1), 105–116.

Kloosterman, R. C., & Musterd, S. (2001). The polycentric urban region: Towards a research agenda. *Urban studies, 38*(4), 623–633.

Kloosterman, R. C., & Lambregts, B. (2001). Clustering of economic activities in polycentric urban regions: The case of the Randstad. *Urban Studies, 38*(4), 717–732.

Lancichinetti, A., & Fortunato, S. (2009). Community detection algorithms: A comparative analysis. *Physical Review E, 80,* 256117.

Leung, I., Hui, P., Liò, P., & Crowcroft, J. (2009). Towards real-time community detection in large networks. *Physical Review E, 79,* 066107.

Li, Z., Zhang, S., Wang, R., Zhang, X., & Chen, L. (2008). Quantitative function for community detection. *Physical Review E, 77,* 036109.

McDonald, J. F., & McMillen, D. P. (1990). Employment subcenters and land values in a polycentric urban area: The case of Chicago. *Environment and Planning A, 22*(12), 1561–1574.

Meijers, E. (2005). Polycentric urban regions and the quest for synergy: Is a network of cities more than the sum of the parts? *Urban studies, 42*(4), 765–781.

Meijers, E. (2008). Measuring polycentricity and its promises. *European Planning Studies, 16*(9), 313–1323.

Meijers, E., & Romein, A. (2003). Realizing potential: Building regional organizing capacity in polycentric urban regions. *European Urban and Regional Studies, 10*(2), 173–186.

Modarres, A. (2011). Polycentricity, commuting pattern, urban form: The case of Southern California. *International Journal of Urban and Regional Research, 35*(6), 1193–1211.

Nan, D. (2014). *Prediction methods of hot region based on complex networks and communities detection*. Master dissertation, Wuhan University of Science and Technology.

Niu, D., Chen, H., Jin, X., & Liu, L. (2013). Complex network community detection algorithm based on core nodes. *Computer Engineering and Design, 34*(12), 4089–4093.

Papadopoulos, S., Kompatsiaris, Y., Vakali, A., & Spyridonos, P. (2012). Community detection in social media. *Data Mining and Knowledge Discovery, 24,* 515–554.

Parr, J. (2004). The Polycentric urban region: A closer inspection. *Regional Studies, 38,* 231–240.

Redfearn, C. L. (2007). The topography of metropolitan employment: Identifying centers of employment in a polycentric urban area. *Journal of Urban Economics, 61,* 519–541.

Roth, C., Kang, S. M., Batty, M., & Barthélemy, M. (2011). Structure of urban movements: Polycentric activity and entangled hierarchical flows. *PLoS ONE, 6*(1), e15923. doi:10.1371/journal.pone.0015923.

Schwanen, T., Dieleman, F. M., & Dijst, M. (2001). Travel behavior in Dutch monocentric and policentric urban systems. *Journal of Transport Geography, 9,* 173–186.

Shi, C., Yan, Z., Cai, Y., & Wu, B. (2012). Multi-objective community detection in complex networks. *Applied Soft Computing, 12,* 850–859.

Vasanen, A. (2013). Spatial integration and functional balance in polycentric urban systems: A multi-scalar approach. *Tijdschrift voor Economische en Sociale Geografie, 104*(4), 410–425.

Wang, L., & Cheng, X. (2015). Dynamic community in online social networks. *Chinese Journal of Computers, 38*(2), 219–237.

Wen, H., & Tao, Y. (2015). Polycentric urban structure and housing price in the transitional China: Evidence from Hangzhou. *Habitat International, 46,* 138–146.

Yang, J., McAuley, J., & Leskovec, J. (2013). Community detection in networks with node attributes. In H. Xiong, G. Karypis, B. Thuraisingham, D. Cook, & X. Wu (Eds.), *2013 IEEE 13th International Conference on Data Mining, 7–10 December, 2013, Dallas, Texas* (pp. 1151–1156). Piscataway, NJ: IEEE Computer Society.

Yin, H. (2014). *Research on local interested community detection in large-scale social network*. Doctoral dissertation, University of Science and Technology of China.

Yue, W., Liu, Y., & Fan, P. (2010). Polycentric urban development: The case of Hangzhou. *Environment and Planning A, 42*(3), 563–577.

Zhou, L., Yan, L., & Shen, X. (2013). Partition method for community structure in complex networks based on edge density. *Computer Applications and Software, 30*(12), 8–11.

Part III
Visualization of Big Geographical Data

Improving GIScience Visualization: Ideas for a New Methodology

Francis Harvey

Introduction

The aim of this paper is to present the foundations for a framework that supports scientific communication and discovery in today's research institutions. Their diversity leads to requirements that GIScience visualization speak without the limits of traditional cartography nor domain-optimized approaches, that provide clear functionality for visualization, but cannot be applied without great effort and reinvention to other domains. This foundation and framework focus on the broadest possible realms: scientific communication and discovery (Goodchild 2011).

Faced with complexity and the limitations of communications models from the 1960s and 1970s, visualization in GIScience and analytical visualization (MacEachren 1995) developed a strong analytical direction, as support for researchers relying on spatial visualization to understand complex processes and situations. The traditional uses of maps in other phases of research received less attention. The potential here remains constrained by the prevalence of conventional approaches that developed in an era dominated by paper maps. These maps, general topographic maps, specialized topographic maps and broad range of thematic maps, were integrated into the research process during studies or training, and changed slowly and slightly in the decades that followed. Many times, with computer-based processing of GI, while we are amply capable of dealing with large data amounts, we end up relying on traditional types of representations. These representations generally developed for maps. They are static and two-dimensional with symbolization allowing for only the constrained representation of a model or data. We must

F. Harvey (✉)
Leibniz Institute for Regional Geography, Leipzig, Germany
e-mail: F_harvey@ifl-leipzig.de

C. Zhou et al. (eds.), *Spatial Data Handling in Big Data Era*,
Advances in Geographic Information Science,
DOI 10.1007/978-981-10-4424-3_9

acknowledge that conventionalism today can still make sense for organizational and economic reasons, yet it is a coin with two sides. The positive side is that it can help facilitate the creation of representations that are more easily understood. The negative side is the limits in understanding and representation that arises in relying on conventions.

This paper lays out a tentative approach as the groundwork to move beyond these limits. The successful connection of geographic representation with cartographic representation comes through applying the transformational approach of GIScience to visualization. This undertaking has to account for known limits in the perception and cognition of symbolization as well. Functional approaches face constraints in accounting for these limits through constraints to the specific data, activities and even institutions used explicitly and implicitly. Many innovative approaches to GI visualization have been published—they are very significant contributions (Andrienko et al. 2003; Roth 2013; Ferster 2012). This paper describes a framework that can adequately harness them to evolving approaches of scientific research.

Theoretical limits precede the practical limits in science and this framework is broad in scope, based clearly on the concept of transformations. Transformations can only account for what we know. This epistemological problem is connecting how we can use GI for discovery to the transformational capacities. David Sinton's framework (1978) provides a constructive way to define the potential geographic representations. Work by Alan MacEachren that extends Jacques Bertin's framework for graphical variables delineates as well as assesses the suitability of these variables for cartographic representation. The visualization framework from (Börner 2012; Börner and Polley 2014; Börner 2015) and MacEachren's graphic variables (MacEachren 1994) can be linked to the abstraction David Sinton provides to develop a linkage between geographic representation and cartographic representation, culminating in a geovisual framework and a process for applying this framework. This framework addresses the multiple dimensions of complexity and implicit limits of conventions to support scientific communication and discovery in a broad range of research institutions.

The next section starts out with a review of the constraints arising in the predominance of static 2D representation and turns to the issues involved in representation of complex patterns and processes. Known solutions are analyzed in terms of their theoretical application of the transformational approach and addressing methodological possibilities of GI-representation that Sinton described. The following section builds on this framework and draws on Börner's framework to structure and systematize the process of connecting geographic representations to visualizations. Coming back to transformational issues, the process is described as series of steps. In the concluding section, I point to key contributions, describe limitations of this conceptual work, and suggest paths for its development.

Challenges and Transformations

This section starts out considering the challenges of using cartographic representations for the complex patterns and processes of GIScience. The potential of visualization approaches using arrows and animation is connected to methodological issues. Transformational approaches (Tobler 1968, 1979) offer a framework for reframing the process. David Sinton's framework (Sinton 1978) for geographical information representation provides a very significant way to reconsider representation and abstractly approach it.

The transformational approach underpins GIScience (Goodchild et al. 1992; Goodchild 1992; Chrisman 1987, 1999, 1998) and fundamentally and principally reconceptualizes the process of making representations for scientific communication to account for the centrality of information in now representing our observations and measurements from the world. These transformations can be modeled algebraically (Kuhn 2012) allowing us to understand how we can convert and manipulate this information to develop better and new understandings of the world. What we are transforming are data that have information because of our knowledge of its informational value (Gleick 2011; Bateson 2000). In terms of visualizations, here we are often constrained by the complexity of visualizations and predominance of conventions. These conventions can help us by guiding us to representations known to scientists and others that they understand more readily due to their familiarity. They can also hinder the presentation of the scientific findings and discoveries by constraining us to old representations.

We should start from the informational approach to geospatial phenomena and remember their data representations can be algebraically processed and transformed. As Fig. 1 shows, the process, in general, is a matter of beginning with observations and measurements in context, creation of data to digitally store the results and transformations of the measurement frameworks and the information content. This distinction reflects the extremely developed state of cartography, due to the centrality of maps for hundreds of years and corresponding role of the profession, but it remains constrained to transformations of information content through classifications and symbolization. Geographic representation as only around a 50–60 year history and thus is much less established, developed, and integrated into diverse practices and institutions of society. The rapid growth of new methods and theories led to a growing gap with established cartographic visualization techniques. They remain dominant for many researchers, even with grave limits for science (Goodchild 2011). As a result, most processing of geographic information generally follows the sequence that Fig. 1 shows, with a visualization, often a map, the final output of the processing.

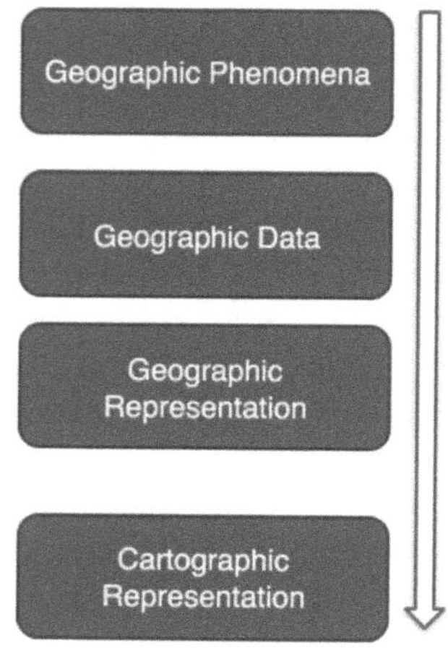

Fig. 1 General scheme for the processing of geographic information

Transformations Implement Operations

Another way to examine the process of transformation is to emphasize the scientific use of operations that implement the transformations. Transformations, following Tobler and others, are that we can algebraically process information representations. Bertins' matrix information processing approach is perhaps the most advanced systematic approach (Bertin 1981). It relies on a system of organizing attributes by geographical entities and then manipulating the matrix for the area of a map with a focus on distinctions in terms of six graphical variables. These graphical variables are translated to retinal variables, which are drawn on the map to ensure the information from the matrix is lucidly communicated. More generically, transformations are understood as way of explaining and understanding how measurements, stored as data with associated geographical entities, are algebraically manipulated to produce new information or new modes of representation, e.g., factoring soil type by slope suitability to determine potential erosion risk or assigning ranges of values to new indicators, which indicate a broader range of phenomena.

Figure 2 illustrates the range of issues and factors that the process of GI transformation takes into account—both implicitly and explicitly. Chrisman's framework explicitly acknowledges the key importance of accuracy in this processing to ensure that results correspond to phenomena in the real world. Adding data visualization aspects and situating the relevant contributions of several authors that this chapter considers graphically shows how visualization transformations actually are connected in multiple ways to geographic representation transformations. Solely those representational transformations of information for a visualization production, resting on cartographic representational concepts, can be separated, although their context is defined by other transformations. The following section will come back to visualization transformations and working from Bertin (1983)

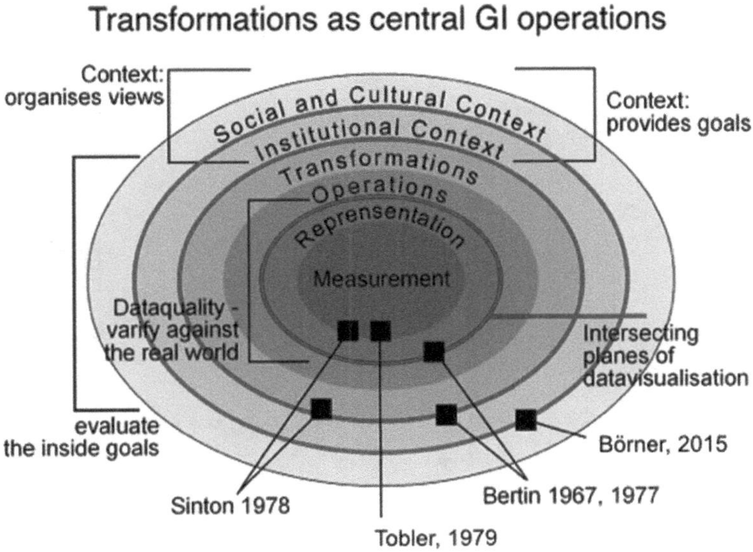

Fig. 2 Chrisman's ring model of GIS activities (redrawn from Chrisman 1987, extended)

and MacEachren's (1994, 1995) contributions place them in a framework that facilitates approaches to GIScience visualization that offers an adequate methodology for the challenges.

The central challenges for GIScience visualization go back to how we can represent phenomena taking place in the world through patterns and processes (Harvey 2016). Understood in the transformational approach presented here, the work with geographical information representations in widely used GIS-software allows us to choose between measurements of attributes organized into a regular tessellation of space, usually a grid, also known as raster, or pre-determining attributes and measuring their spatial extent. Figure 3 provides an example for the distinction. This portion of the world is characterized by different physical features, related to ecological processes, geology, and other influences, e.g., here the organized grazing of animals. Some processes lack a representation here. This is because of choices made in prioritizing data collection for specific purposes arising from the institutional and social and cultural contexts. In any case, the observations and measurements can be represented following either a vector coverage model or a raster data set. The vector coverage model, following Sinton, measures space based on fixed attribute characteristics, e.g., the type of vegetation or land cover. The raster data set fixes space into a grid with a specified resolution and based on an analysis of measurements and observations in each cell assigns an attribute. In both approaches that Sinton (1978) describes, time is controlled. The observations and measurements were collected in a controlled time frame, e.g., the first two weeks following on the spring equinox or the date and time when the satellite sensor recorded the data. This means in this example that processes have already been

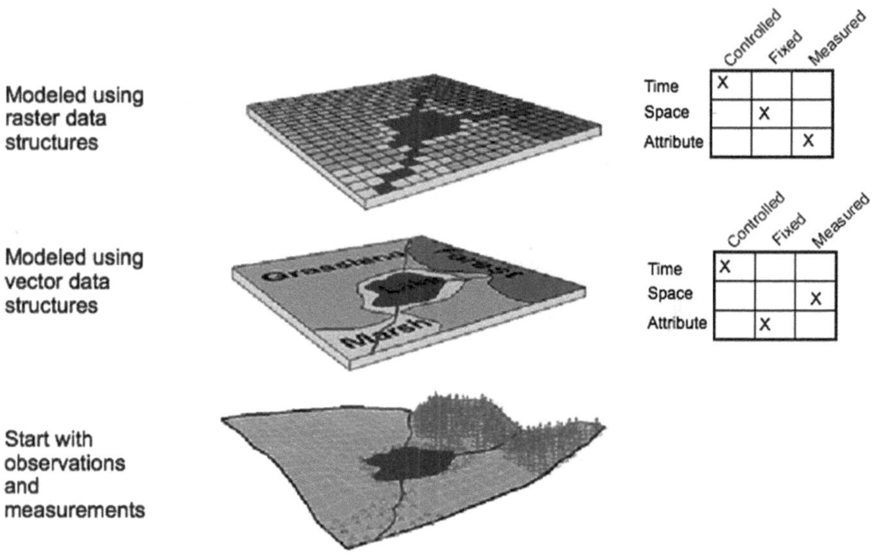

Fig. 3 Sinton's framework and vector coverage and raster models in comparison (from Harvey 2016, used with permission)

accounted for in data collection. For example, a field biologist could sample vegetation in distinct areas and assign that area a type of land cover classification. This can be quite detailed and take into account flooding, succession processes, anthropogenic influences. The resulting information is of patterns that only biological and ecological knowledge in this example could interpret to develop an understanding of processes. Through systematic description of processes it becomes possible in this approach to address fundamental questions about phenomena, semantics, motivations and transformations. Information about these aspects can be related to questions about data, accuracy, geographical representation, semiotics, and map elements.

With this knowledge, which can also be partial and incomplete, it becomes possible to reliably transform data and conduct spatial analysis. Depending on the level of knowledge and other contingencies, an analyst may use various GIS-operations (overlay or buffer, for example) to transform the data and assess influences and processes through visualization of the results. This processing is the heart of GI analysis and representation, but with many unknown contingencies cannot be algebraically represented it is a process of communication with many complexities. The next section goes on to consider these limits and describe a heuristic approach to better account and more reliably transform and represent GI. This also supports the iterative development and testing of symbolization.

A Conceptual Approach and Tentative Framework

This section builds on the concepts presented in the previous section and describes a heuristic approach based on transformations and involving a systematic sequence of steps and questions to guide the transformations phenomena, semantics, and motivations. Issues related to data, accuracy, geographical representation, semiotics, and map elements also guide the process which culminates in an iterative selection of symbolization. Katy Börner's framework structures the process of connecting geographic representations to visualizations as a systematic process of transformations.

Beforehand though, given the increasing importance of visualizations in scientific research, its changing nature leads to evolving potential for integrating visualization in various stages of research and implement it iteratively to support a broad range of potential activities. This change also reflects increased interaction possibilities with visualization and the greatly improved ease for creating visualizations. Finally, data-centered research (Bell and Gray 1997; Wing 2006) opens up potential for these types of visualization to be integrated throughout the research process. The ideas the Edward Hutchins in Cognition in the Wild bear consideration here, as scientific research becomes less and less hierarchical. From this book, and his analysis of complex decision making activities that constantly require engagement with situations as they change, we can describe a cycle of measurement, computations and interpretation that is repeated in the scientific research process. For visualization, this means a more tightly, yet in ways that depend on research organization, more flexible approach to connecting transformations and visualizations.

In Hutchin's analysis when computation through coordination becomes more central to the process, successful communication means successful interaction. In his study, he focuses on the situational development of specific languages to enable interactions as required, considering how the coordination of individual actions involves resolving relationships with resources and constraints in an organization. People are always unraveling this complexity. Communication in this sense is interaction. It is an interaction that strongly relies on visualizations. Visualization integrated in this way as communication also offers a heuristic approach for knowledge discovery.

Specifically, considerations of scientific visualization have to connect the transformations that take place with the interactions they are connected to. Together they make up the scientific visualization process. These aspects should start, following Hutchins, with seeing—thinking—acting as the three analytical phases always involved in visualization. In these phases, people connect semiotics with semantics and either create the representations or understand the representations in context. Hutchins work focuses on how people use artifacts as cognitive extensions, not whether they are producers or users. In this sense, Hutchins returns to the original breadth of the communication model. Accounting for the processes through which people reduce complexity and deploy or follow conventions also considers the role of Gestalt psychological rules. Finally, assessing the efficiency of the interactions should also account for understanding of representational accuracy. It may be readily possible to quickly grasp a scientific visualization, but understanding its accuracy may be far more challenging and far more fundamental to the research.

Katy Börner's scientific visualization framework offers a robust structure for integrating visualization-related activities into these changing research workflows and scientific communication (Börner and Polley 2014). As Fig. 4 shows, the workflow consists of several steps in an iterative loop. Starting with context, explicitly for Börner defined as stakeholders, and important parameters for the scientific visualization, begins with reading, then analyzing and afterwards visualizing the data through selections, combinations using overlay and selection of the visual encoding, or symbolization. Börner's framework is much broader and designed for all types of visualization. With the focus here on GI representation her framework offers a set of steps that can be developed iteratively, to more closely link visualization with scientific communication.

A heuristic approach this chapter suggests involves extending Börner's framework to explicitly integrate GI transformations at all stages. The graphic can only suggest the range of possibilities. Turning back to the example of land cover from before, a more exhaustive list of possible transformations can be drawn up. A full list can be derived from Chrisman's publications on this topic (Table 1).

Since Börner's framework creates an iterative loop, processes of quality assessment, refinement can be implemented and guide the transformations. Theoretically, whether it is possible to return to the original data, either to assess previous transformations, add additional insights and variables, or to even start anew, depends on the level of validation, interpretation or clarification desired. Transformations can alter the semantics of data and without this recursion,

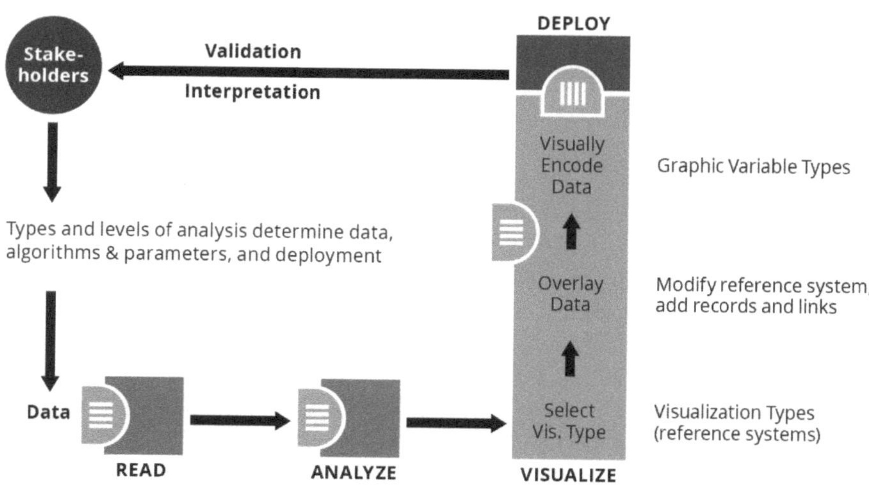

Fig. 4 Needs driven workflow design (based on Börner and Polley 2014, used with permission)

Table 1 Possible Transformations in exemplary GI-processing

Step	Possible transformation
Read	Recode
	Simplify/reclass
	Combine
Analyze	Modify
	Reclass
	Composite
	Integrate
	Buffer
Visualize	Recode
	Simplify
	Encode

it is conceivable and possible that some users will lack understanding of the data and potentially make erroneous interpretations. The iterative cycle thus offers the potential to ensure accuracy and reflect also on motivations. In regards to visualization, the documentation and understanding of the processing is key to connecting semiotics and semantics. For validation and interpretation, but also during analysis, the understanding of what symbols show and how they correspond to phenomena. This is the basis for reviewing and assessing the accuracy of transformations. This knowledge encourages a reflective and possibly iterative exploration of possible visualization symbology. The processing can be described in a flow-chart or model to assure reliable communication to others. This, of course, is essential for creating visualizations with adequate spatial and temporal depth to adequately show the geographic complexity and how the patterns and processes of the visualization aid our understanding.

Conclusion: A Tentative Foundation

This paper presents the tentative foundation for a framework that supports scientific communication and discovery in today's research institutions. It provides an initial framework for geographical information visualization. It focuses here on the conceptual level. Clearly further work refining the framework and assessing it are called for. The key contributions of the framework is that it addresses the challenges and complexity of geographic information visualization in diverse settings and the possibilities of moving on from conventional static 2D representations to visualizations with adequate spatial and temporal depth to show this complexity. In contrast to constraints of paper-based media, contemporary and future visualizations can use any conceivable digital or analogy means of representation. More important for the sciences, constrained by the economic constraints of established forms and institutions for publication and dissemination, the variety of scientific research approaches means the diversity of visualization approaches calls for diversity in breadth and depth. The discursive approach based on Hutchins' work outlined here offers both. The clear need to move beyond static and 2D representations to adequately visualize patterns and processes of geographic phenomena requires a different approach to visualization than traditional modes of cartography. Yet, this framework, as suggested, can also be used to create traditional cartographic representations—which have the advantage of being close to wide-spread conventional approaches and thus be easier for people to visually comprehend. Scientific research benefits from the new possibilities that a broader approach to visualization offers.

The limits of this preliminary work lie predominantly in the conceptual orientation of this work. While it uses examples to provide illustrative descriptions, a proof of concept, refinements and application are clearly called for. It is also constrained by only considering possibilities to implement this approach in current GIS software architectures and their raster and vector data structures, both static and two-dimensional.

Future work on this approach should also focus on enhancing our understanding of the connections between semiology and semantics that are central to scientific visualization. In this sense, consideration of the changing roles of people and institutions involved in scientific research takes on great significance for both understanding its changing roles and improving it in GIScience.

References

Andrienko, N., Andrienko, G., & Gatalsky, P. (2003). Exploratory spatio-temporal visualization: An analytical review. *Journal of Visual Languages & Computing, 14,* 503–541.
Bateson, G. (2000). *Steps to an ecology of mind.* Chicago: University of Chicago Press.
Bell, G., & Gray, J. (1997). The revolution yet to happen. In P. J. Denning & R. H. Metcalfe (Eds.), *Beyond calculation: The next fifty years of computing.* New York: Springer Verlag.
Bertin, J. (1981). *Graphics and graphic information processing.* Berlin New York: de Gruyter.
Bertin, J. (1983). *Semiology of graphics: Diagrams, networks, maps.* Madison, WI: University of Wisconsin Press.
Börner, K. (2012). *Places and spaces: Mapping science.* Cambridge, MA: MIT Press.
Börner, K. (2015). *Atlas of knowledge: Anyone can map.* Cambridge, Massachusetts: The MIT Press.
Börner, K., & Polley, D. E. (2014). *Visual insights: A practical guide to making sense of data.* Cambridge, MA: The MIT Press.
Chrisman, N. R. (1987). Fundamental principles of geographic information systems. In N. R. Chrisman (Eds), *Auto-Carto 8, ASPRS* (pp. 32–41).
Chrisman, N. R. (1999). A transformational approach to GIS operations. *International Journal of Geographical Information Science, 13*(7), 617–637.
Chrisman, N. R. (1998). Rethinking levels of measurement for cartography. *Cartography and Geographic Information Systems, 25,* 231–242.
Ferster, B. (2012). *Interactive visualization: Insight through Inquiry.* Cambridge: The MIT Press.
Gleick, J. (2011). *The information: A history, a theory, a flood.* New York: Vintage Books.

Goodchild, M. (2011). Challenges in geographical information science. *Proceedings of the Royal Society A, 467,* 2431–2443.
Goodchild, M. F., Haining, R., & Wise, S. (1992). Integrating GIS and spatial data analysis: Problems and possibilities. *International Journal of Geographical Information Systems, 6*(5), 407–423.
Goodchild, M. F. (1992). Geographical information science. *International Journal of Geographic Systems, 6*(1), 35–42.
Harvey, F. (2016). *A primer of GIS: Fundamental geographic and cartographic concepts, 2nd ed.* New York: Guilford.
Kuhn, W. (2012). Core concepts of spatial information for transdisciplinary research. *International Journal of Geographical Information Science, 26*(12), 2267–2276.
MacEachren, A. M. (1994). Some truth with maps: A primer on symbolization and design. Washington D. C.: American Association of Geographers.
MacEachren, A. M. (1995). *How maps work: Representation, visualization, design.* New York: The Guildford Press.
Roth, R. E. (2013). Interactive maps: What we know and what we need to know. *Journal of Spatial Information Science, 6,* 59–115.
Sinton, D. F. (1978). The inherent structure of information as a constraint to analysis: Mapped thematic data as a case study. *Harvard Papers on Geographic Information Systems, 7,* 1–17.
Tobler, W. (1968). Transformations. In J. D. Nystuen (Ed.), *The philosophy of maps.* Ann Arbor: University of Michigan.
Tobler, W. (1979). A transformational view of cartography. *The American Cartographer, 6,* 101–106.
Wing, J. M. (2006). Computational thinking. *Communications of the ACM, 49,* 33–35.

Leveraging Big (Geo) Data with (Geo) Visual Analytics: Place as the Next Frontier

Alan M. MacEachren

Introduction

Place matters. It is a fundamental component of everyday life and has been a core topic of Geography since Aristotle (Morison 2002). GIScience, however, has directed much more attention to "space" than to "place" in its approaches to information collection, organization, analysis, and decision-support. This focus on formal approaches to space and precise location specification has served GIScience and related geographical information technology developments well, in leveraging the dramatic increases in geo-referenced data for a wide range of applications, But, the lack of attention to place has created a gap between the methods and tools now available and the needs of science and society for multifaceted understanding of the world. Plus, lack of attention to non-traditional geospatial data has left a vast resource of untapped place-relevant data that is unstructured and thus not accessible by current spatial database, analytical, and other tools.

Big Data matter—they have the potential to enable GIScience to move beyond the current spatial focus to address scientifically and societally important questions of place. It has been nearly a decade since the term "Big Data" gained prominence as a popular label for the challenge | opportunity that our instrumented world is prompting | providing. Big Data is a somewhat misleading term, since the concept is typically characterized as being about more than just data size (Volume). It is also about Velocity (the speed at which new data arrives) and Variety (the heterogeneity in type, quality, and other characteristics); and as outlined below, some argue for even more components.

A.M. MacEachren (✉)
GeoVISTA Center, Department of Geography, The Pennsylvania State University, University Park, PA 16802, USA
e-mail: maceachren@psu.edu

C. Zhou et al. (eds.), *Spatial Data Handling in Big Data Era*,
Advances in Geographic Information Science,
DOI 10.1007/978-981-10-4424-3_10

Government agencies, businesses, and other organizations are gearing up to meet the Big Data challenge | opportunity with strategies to both generate and leverage Big Data; and science funding organizations have initiated a range of calls for Big Data research. For Geographical/Spatial Information Science, the big data challenges and opportunities require a fundamentally new perspective on geospatial data, one that treats geospatial data as an integral component of an information ecosystem in which (geo)spatial may still be special, but in which geospatial data cannot be considered independently from other kinds of data, … or from the context of use, … or from the knowledge and needs of users.

The argument presented in this paper is that the advent of big geospatial data (and development of data science methods to leverage those data through connections to other data, context of use, and knowledge/needs of users) offers an opportunity to address questions of place in fundamentally new ways. The view presented focuses not on formalizing place in ways that support application of existing modeling and analysis methods from GIScience, but on embracing the complexity inherent in conceptions of place as socially constructed and imprecisely delineated entities and leveraging advances in data availability and methods to explore that complexity.

Many industry estimates suggest that 80% (or more) of big data are unstructured (Andriole 2015). Much of these data are likely to contain some form of geographical reference through place names, descriptions of geographic-scale events and behavior in places, and other relative indications of location. But, that geographically relevant data is often ignored because our existing methods and tools take a 'space' focus while much unstructured data reference 'place' through natural language. Thus, understanding place, as it is reflected in language (as both an entity talked about and as a context within which described events and behavior happen), is a fundamental question in Geography and essential to developing geographical information retrieval (GIR) methods that leverage the wealth of geographical data embedded in text and related unstructured sources. In complementary fashion, developing geographical methods and tools for retrieving place-based information from unstructured text data sources and enabling users to leverage that information together with more traditional data sources offers new opportunities to connect place and space.

This paper presents an argument for a "human-in-the-loop" (geo)Visual Analytics approach to leveraging big (geo)data as a means to understand and enable experience of place as a dynamic construct. A focus is put on leveraging unstructured geographical data found in text sources. The approach contrasts with those that rely exclusively on computational methods to produce information and generate answers. (geo)Visual Analytics (gVA) is presented as both a science and set of methods/tools that are focused specifically on support of human analytical reasoning with big, heterogeneous, dynamically changing, and often 'messy' data that include geographical components. And, an argument is presented that gVA applied to these rich and dynamic data offers new windows to understanding place in ways that are not provided through traditional Geographical Information System (GISystem) methods applied to structured geospatial data.

Place: A Snapshot

Place is a complex concept that it is impractical to discuss in depth here. For those interested in in understanding the concept more fully, a comprehensive introduction is provided by Cresswell (2014). Some additional book length treatments of place from a social science and humanities perspective include: Agnew and Duncan 2014; Carmona 2003; De Blij 2008; Duncan and Ley 1993; Ellard 2015; Hubbard and Kitchin 2010; Massey 2013; Nairn et al. 2016; Relph 1976; Tuan 1977. Here, I will sketch just an outline of some of the complex issues about place for which big data have a potential to enable insight.

An important starting point to understand place as distinct from space, is Agnew's (2011) direct analysis of the distinctions and interrelationships of these two fundamental geographical concepts. Agnew presents space as the more abstract concept, grounded in conceptions of location (both absolute and relative) and reflected in twentieth century perspectives of "spatial science". Place, in contrast underlies conceptualizations of geography as a "science of places", with a holistic approach to places as dynamic, thus defined by activities and processes. Prototypical of this view is Pred's (1984) argument for place as a complex time-space activity, replete with power relations, culture forms, biographics, and relationships to nature.

With this context, Agnew (2011) makes important distinctions about how "place" is conceptualized, either as location (thus "assimilated to space") or as occupation of location. He elaborates on this distinction by characterizing the location view "as nodes in space simply reflective of the spatial imprint of universal physical, social or economic processes" (thus a "mere part of space") and the occupation view "as milieux that exercise a mediating role on physical, social and economic processes and thus affect how such processes operate" (thus "a phenomenological understanding of a place as a distinctive coming together in space").

Even with the place as location conceptualization, however, Agnew (2011) points to the dynamic and interconnected nature of places. Specifically, he extends from his initial location-occupation distinction to define three 'dimensions' along which the meaning of place is defined within the various theoretical positions from which place is considered. The first corresponds to the place as location view, specifically the meaning of place along this dimension is characterized as a "… location or a site in space where an activity or object is located and which relates to other sites or locations because of interaction, movement and diffusion between them." Then, the occupation view is further parsed into two additional meaning dimensions.

The second dimension characterizes "…place as a series of locales or settings were everyday-life activities take place. Here the location is not just the mere address but where of social life and environmental transformation" (Agnew 2011). These locales provide the social setting of everyday life that can include workplaces, churches, schools, etc., but also non-fixed settings of activity, such as vehicles or chat rooms. As noted by Cresswell, one mechanism through which spaces can become places in this sense is through naming. Places of importance, due to activities that they support, are given names while 'spaces' that do not meet a need or support recurrent behavior are not (and thus do not become) places.

The meanings associated with the activity-based places vary with geographical scale. Small places have meanings related to self while big places have meanings associated with others or with the environment (Gustafson 2001). Massey's (1994) perspectives on place seem relevant to this dimension of place meaning. In particular, she critiques a common view of place as "bounded entities" (with inside clearly distinguished from outside) and with single, essential entities. Places, according to Massey should always be regarded in relation to the outside world. Places can be special due to linkages to the outside world (rather than their own intrinsic qualities). Massey (1994, p. 154) argues that places "...can be imagined as articulated moments in networks of social relations and understandings, but where a large proportion of those relations, experiences and understandings are constructed on a far larger scale than what we happen to define for that moment as the place itself, whether that be a street, or a region or even a continent."

The third dimension of place meaning focuses on "... place as sense of place or identification with a place as a unique community, landscape, and moral order" (Agnew 2011). This latter view might be thought of as the humanistic conceptualization of place in contrast to the more social science perspectives of the first two dimensions. Representation of place along this sense of place dimension is typically verbal or visual. Coordinates are not place, but a description or photos of what is near them can invoke a sense of place. But, from this sense of place perspective, the actions of individuals in 'creating' the place through various activities, particularly those that may be 'unofficial' is part of what generates a rich sense of place. One intersection between GIScience and sense of place is, perhaps, the many volunteered geographic information (VGI) activities that citizens are engaging in (Hardy et al. 2012). One example is a recent project by Quinn and Yapa (2015) to help communities in Philadelphia create greater food security by mapping the informal food resources in their communities (e.g., urban gardens, sources of compost or organic matter to support those gardens, farmer's markets, food banks and soup kitchens).

Vasardani and Winter (2016), considering place from a GIScience perspective, argue that place "...is a location (in an environment, not in an empty space) with properties that give it 'shape and character' and which enable conversations about place." Their perspective draws upon the "theory of centers" from architecture. Grounded in this theory are 15 structural properties proposed by Alexander (2002). An argument is made that having a 'center' and a gradient away from center is fundamental to places and that places are seldom considered independently of other places and relationships; thus there is an emphasis on interconnectedness that reflects the social science arguments outlined above.

Big Data

Place has been a core concept of Geography for centuries, but one that has been difficult to formalize sufficiently in order to leverage digital data to support understanding of place as a dynamic construct. The structured digital data so well suited to supporting spatial analysis have been an impediment to analysis of place since they separate location from meaning. But, the advent of Big Data is creating a context within which new data-driven approaches to under-standing place may become possible. We now have: (a) an abundance of geo-located (or geo-locatable) data that serve as an input to geo-analytical reasoning and (b) many new map-based and other visual methods and technologies that purport to help people reason with and make decisions based upon these big data. To take advantage of these developments to address questions of place, we need to consider both the challenges and the opportunities that big data provide. In particular, I draw upon a characterization of the "5 Vs" of big data by Monroe (2013). They are:

- *Volume*: massive data scale due to sensors, electronic transactions and records, ubiquitous data generation via smarts phones & social media;
- *Velocity*: rapid data update due to continuously operating sensors and data generators + streaming technology;
- *Variety*: heterogeneity in types of data, many of them never before seen;
- *Validity*: varied and uncertain reliability of the data, its processing, its interpretation, and resulting decisions; a key is construct validity: the degree to which the technique measures what it claims to be measuring;
- *Vinculation*: to "vinculate" is to bind together, to attach in a relationship; it is about what might be described as the fundamental interconnectedness of all things (Richardson et al. 2012).

All of these components of big data are relevant to understanding place and to leveraging place-relevant data to support scientific and societal challenges. The fifth, vinculation is particularly relevant to the potential for leveraging big data to understand place and to conceptualizing place in an era of big data.

Place (from a theoretical, geographical perspective as outlined above) is an "experience-based dynamic construct" (Agnew 2011). Connectedness of places is also inherently dynamic, thus, big, streaming place-linked data, for the first time, make it possible to develop methods allowing insight into the geo-social dynamics of places and their massively interconnected, changing nature. But, if we are to leverage information about place from these largely unstructured and semi-structured data, we need to develop a better conceptual model of how place is signified in language (particularly in text) in order to recognize, retrieve, and analyze the wealth of references to place that have gone mostly untapped thus far. While more than a decade of research in GIR has made important progress, most of that work has focused only on the problem of recognizing and geolocating place names, thus turning place into space and/or on linking documents to a geo-graphic "footprint" for which they are determined to be relevant (again turning place into space to support integration of data derived from text into traditional spatial analysis). This space-centric work needs to be complemented by developing a rich characterization of what it means for a document to be "about" a place and how to recognize and interpret statements about place that lack formal place names. Here gVA is proposed as a method and suite of tools that can help achieve this objective.

(Geo)Visual Analytics

"Visual analytics is the science of analytical reasoning facilitated by interactive visual interfaces" (Thomas and Cook 2005). The initial focus of efforts in the field was on visual-computational support for assembling evidence, generating inferences and explanations from evidence, comparing /assessing those inferences and explanations, and reporting results (e.g., Andrienko et al. 2011; Keim et al. 2010; Kohlhammer et al. 2009; Malik et al. 2012; Robinson 2011; Tomaszewski and MacEachren 2012; Wang et al. 2008). As the field has developed, increased attention has been directed to big data (e.g., Andrienko et al. 2013; Keim et al. 2013). A 2010 Visual Analytics research agenda report from Europe proposed that "Visual analytics combines automated analysis techniques with interactive visualizations for effective understanding, reasoning and decision making on the basis of very large and complex datasets" (Keim et al. 2010). Building on these ideas, and focusing on geographical big data, I offer the following definition of gVA: *Geovisual Analytics is a domain of research and practice focused on visual interfaces to analytical methods that support reasoning with and about big, dynamic, heterogeneous, unconfirmed, hyper-connected geo-information—to enable insights and decisions about something for which place matters.*

The five "V"s of big data are reflected in the qualifiers on geo-information in the definition above as is the focus here on going beyond traditional spatial analysis to consider place. Since visual analytics was identified as a specific domain of research and practice, more than a decade ago, substantial progress has been made on developing new computational methods to deal with the challenges of big data and on visual interface methods to couple human knowledge, reasoning, and insight with these computational methods. But, although geolocated data is often a focus of these efforts (by GIScientists and others), place (in contrast to space) has been given only limited direct attention. The remainder of this essay presents an argument for, and selected early steps toward, taking advantage of advances in visual analytics, coupled with big data (in all its guises) to address place as a subject of specific attention.

Leveraging Unstructured Big Data to Understand Place: Taking a gVA Approach

There is a long history of spatial science and technology, both within geography and more broadly across the disciplines that coordinate research under the umbrella of GIScience. That history includes fundamental advances in how we collect spatial data and assess its fitness for use (e.g., Bharti et al. 2011; Martin 1998; Woodcock and Strahler 1987); in how we represent, store, and retrieve those data (e.g., Langran 1992; Mennis et al. 2000; Peuquet 1988); in spatial analytical methods (e.g., Anselin 1995; Charlton et al. 2006; Hubert et al. 1981); and in qualitative spatial reasoning (e.g., Egenhofer and Herring 1990; Klippel et al. 2012). In contrast to the focus on space, there is a very short history in GIScience (or gVA) of research directed to place (for a few examples, see: Agarwal 2005; Bennett and Agarwal 2007; Edwardes and Purves 2007). The relatively recent argument by Goodchild (2011) that attention is needed to formalize concepts of place for integration into work with GISystems has prompted some recent attention (nearly 50 citations as of October, 2016, e.g., Roche and Rajabifard 2012; Scheider and Janowicz 2014; Winter and Freksa 2012; Yang et al. 2016). But, only a very small proportion of that work addresses place in the rich sense that it is generally considered by human geographers, other social scientists and planners, or humanists.

An opportunity exists through the advent of big data, to address the nearly infinite complexity of place and its multifaceted connectedness. The challenge that must be met in order to take advantage of this opportunity is to develop strategies and methods to capture and reason about human concerns with place that are potentially represented within the complexity of big data.

My contention here is that three of the five "V"s are particularly central to moving attention in GIScience from 'space' to 'place'. Place is a dynamic construct, thus data *velocity* increases that are associated with big data advances can enhance the granularity with which place can be understood and streaming data on its own (whether high velocity or not) provides a direct window on the dynamic nature of place. Place is also multifacected with multiple layers of embedded meaning. Data *variety*, therefore is an essential input to understanding the multifaceted structure of place. Place is also embedded in the context of the world and its diverse connections and complex relationships among entities; place is probably best understood as being hyper-connected. Thus, *vinculation*, with its focus on connection and relationships, complements data velocity and variety as a core component of the data needed to understand and enable behavior in place. Below, I sketch a few initial ideas about each of these aspects of big data, from the perspective of the role that gVA can play in leveraging big data to construct meaningful information about and enable activity in place.

Velocity: Dynamic Data to Represent a Dynamic World

Streaming data are changing the landscape of information technology and decision-making, with impacts across business, government, and science (Madia 2015). The new and rapidly increasing sources of streaming data, much with some kind of geolocation (or potential for geolocation through mechanisms such as geoparsing of place references in text), generates many possibilities for GIScience to consider place in new ways. The human-in-the-loop approach of gVA is well suited to leveraging large, complex streaming data to achieve insights about place, which is itself dynamic as outline above. The computational methods of gVA are needed to cope with the data flow and the visual interface to those methods is needed to both interpret output from the computational methods and steer the methods to cope with changes in data content and form over time, as well as changes in kinds of insights about place needed in a changing world.

More specifically, streaming data provide a key opportunity specific to understanding place because place (from a theoretical, geographical perspective) can only be understood through attention to the dynamic process of activities and events from which places are constituted. Integration of multiple sources and forms of streaming, place-linked data offers the opportunity to observe and analyze the dynamic processes and activities associated with a place. Connectedness of places is also inherently dynamic; across 4 theoretical perspectives on place, Agnew (2011) cites a "... stress on the fluidity and dynamic character of places as they respond to interconnections with other places." Thus, big, streaming place-linked data, for the first time, make it possible to develop methods allowing insight into the geo-social dynamics of places and their hyper-connected, changing nature.

Initial examples of the ways in which these new sources of streaming data can be used to develop a deeper understanding of places include research to leverage cell phone data (e.g., Ratti et al. 2010; Xu et al. 2016), Twitter (e.g., Jenkins et al. 2016; Wojcik et al. 2015), and photo sharing sites (e.g., Andrienko et al. 2015; Liu et al. 2015). As discussed above, places are created, exist, and change continuously as a result of human activity. Research by Kraft et al. (2013), as one example, demonstrates the potential of leveraging unstructured streaming text (from Twitter) within a gVA application to identify the dynamic creation and evolution of an informal place. They provide a use case example of their application in which an analyst is able to recognize a situation where political tensions led to a riot, thus an informal place was generated, and then through social media this informal place was connected across the globe to other places in which related events happened.

Data Variety

As outlined above, place is dynamic and multifaceted, a concept conceptualized as having multiple dimensions. Thus, while data variety is a technological big data challenge, it is also essential to a rich characterization and understanding of places and to support for activities in those places. The variety of data needed to address questions of place goes well beyond traditional spatial data that existing GIScience methods and tools have been designed to collect, store, and process. As outlined in the introduction, unstructured data (in the forms of text, images, and video), are being generated at rapidly increasing rates and much of those data contain at least indirect reference to places. The unstructured data offer an important complement to traditional quantitative geospatial data that is critical to questions about meaning of place.

Here, I highlight two exemplar *data variety* foci that are central to developing new GIScience/gVA methods that can identify, characterize, and support understanding of place: (a) text analytics—extracting place references and meaningful information about place from unstructured, often fragmentary data from a wide variety of sources; and (b) "extreme" information fusion—constructing place characterizations through integration of structured and unstructured data.

Text Analytics

There is probably much more place-relevant data locked up in text data sources than in all forms of traditional geospatial databases. For example, we have found that more than half of all Twitter tweets have some form of place reference (which includes a location that the tweet is from, places mentioned in the tweet text, and/or places that the Twitter user specifies in their profile) (Pezanowski et al., submitted). Similarly, virtually all news stories have a location where the story was posted and also often mention places in the text of the story, particularly in any story about events.

An example of (partially) understanding the dynamic complexity of a place through gVA text analytics methods is provided in SensePlace, one of the first gVA tools developed in our research group specifically to leverage text data sources (Tomaszewski et al. 2011). SensePlace was built specifically to support document foraging and sensemaking designed to understand the seasonally dynamic nature of regional and national population patterns in Niger (as input to infectious disease modeling). Specifically, SensePlace enabled analysts to 'mine' a news archive in order to achieve multiple linked objectives: (1) see where events are and how they relate; (2) know when events happened; (3) visualize links between map/article/timeline/concepts; (4) explore multiple search strings together; (5) save searches and share; (6) focus on relevant documents by eliminating less relevant articles. A core capability of SensePlace was a set of computational methods that recognize and geolocate place names. As noted above, one thing that distinguishes 'places' from 'spaces' is that the former are given names due to their importance. Thus, recognizing and geolocating place names in text is a key step in the process of turning unstructured text into place-relevant data. But, it is a step that is both challenging to do (see Table 1) and that only partially captures the place references in text; recognizing and locating place description that does not include proper place names is an open problem.

Beyond determining the "where" of entities or statements in text, substantial progress has been made in natural language processing that is relevant to the "what" and "why" aspects of places. For example, Nelson et al. (2015) use computational methods integrated into a web-based gVA application to explore differences of opinion about political situations in the U.S. by Congressional district. While this research used existing Congressional boundaries as 'bins' into which twitter data were aggregated and opinions rated computationally, the methods could easily be extended to support identification of places with shared opinions. In related work, Liu et al. (2015) demonstrate methods that extract place semantics from photo tags, providing a means to characterize the sense of a city (using Paris as an example). They go on to propose a comprehensive approach to "social sensing" that complements ideas below on information fusion.

In addition to assessing the "sense" of statements (e.g., opinion, sentiment), a range of methods for topic modeling have been develop that computationally identify sets of discourse having semantic/thematic similarity. One recent example that applies these methods directly to deriving a "sense of place" from text sources is reported by Jenkins et al. (2016). These authors focus in particular on investigating the scale of places and find that (at least for Twitter and Wikipedia) particularities of places can be derived at neighborhood levels but that analysis at city scale provides more insight about the differences between text media than it does about the unique features of places. This finding relates to the above discussion of scale-dependent meanings associated with activity-based places.

Extreme Information Fusion

As noted in the snapshot on place (Section "Place: A Snapshot" above), place as conceptualized in human geography and other social sciences, is a complex concept that is multifaceted, dynamic, and with flexible geographical bounds. Characterizing place, thus requires both the application of multiple perspectives and the integration of multiple kinds of data. The challenge is what I label as *extreme information fusion*, a term intended to characterize the scale of data, the multiple kinds of data, the continually changing nature of data, and the need to build connections across data that are all needed to represent places. While few attempts have been made to apply information fusion methods to understanding place, there are advances in this domain that are relevant and that have the potential to be repurposed to focus more directly on place.

One exemplar is recent work by Cervone et al. (2016) focused on leveraging heterogeneous data in support of crisis response. The authors illustrate how multiple traditional geospatial data sources can be combined with novel unstructured data sources to better characterize events in places in order to support crisis response.

Table 1 Place entity recognition is challenging due to the variability and imprecision of natural language. Below are a few representative examples of tweets containing place names with non-locational, ambiguous, or vague senses; the kinds of references that are challenging to process automatically

Examples of tweets with 'place names' used in ways other than to signify a place
• Noun adjuncts: here place names are used, not to specify the location but as a modifier of another noun, often a person or an organization; there are several variants, as illustrated, that need to be addressed differently by computational methods designed to decide when a statement is "about" a place – Qualifying/naming an event RT @miamivice_22: It is a photo at the time of the Great Hanshin-Awaji Earthquake. Picture hell. http://t.co/rwzzhexLgn – Qualifying a person: RT @ezralevant: Watch the riot videos. Listen to the victims of this violence. And then help me hire Calgary's best lawyer to sue the offen – Qualifying a more precise generic place: Iran: protest rally in front of Gilan governor office against the shutdown of Looshan Cemnet Factory http://t.co/8wkH2239CH – Metonymy: RT @Watcherone: South Sudan rebels have killed several Uganda soldiers in the Upper Nile in renewed fighting in the country – Regular polesymy: RT @we_support_PTI: All Pakistani's In USA - COME OUT to Protest #GoNawazGo, as the #FakePrimeMinister visits United Nations. #ImranKhan ht…
• Ambiguous: places are often contained within other places with the same name (as with Gaza, the city, that is within Gaza, the territory) – RT @saidshouib: #Gaza_Under_Attack | This is not the effect of an #earthquake, also it's not a #meteor. It's an Israeli Missile. http://t.c…
• Vague spatially: informal places are often described relative to formal places – Breaking M6.0 earthquake jolted the sea area near S. Sumatra Wed., the quake hit at a depth of 10 km.(CENC) http://t.co/E0NkG1mbkV
• Vague meaning of place: the meaning of a "place" depends on ones experience with that place; the two individuals posting the tweets below are likely to have extremely different conceptions of Beijing as a place – Getting ready to see @ladygaga!!!!!! I am BEYOND excited!… Back in HK but still high from the thrill of visiting Beijing. Here is me and the Great Wall. Yes – The latest Tweets from Amy Mathieson (@AmyWMathieson). Western trained architect living in Beijing and following building restoration, eco-tourism, and…

Specifically, they fuse multiple data sources over the cities of Boulder and Longmont, CO including: (a) Flickr ground photographs, (b) tweets, (c) Civil Air Patrol images, (d) Falcon UAV, and the (e) satellite water classification. The fusion of this information is shown to support accurate predictions about road closures in specific places. The process enables dynamic update of the continually changing nature of the places impacted by a natural disaster. In this case, the focus is on flooding, but the methodology would support understanding of the dynamic situation in particularly places as any kind of natural hazards or other emergency events evolve.

Vinculation: Extreme Connectedness

Perhaps the most important qualitative change that big data brings over traditional geospatial data is that related to vinculation, the inter-connectedness of all things. Prior data sources tended to put data into 'silos' by type, making it difficult to identify and leverage connections among disparate kinds of entity. But, as noted above, place is fundamentally interconnected at multiple scales. To investigate and understand place at a substantive level requires data and methods that can cope with and leverage the connections. As data about connections becomes increasingly available, geographers and others have begun to explore those connections. In one representative study, about the *Geography of talk in Great Britain*, Ratti et al. (2010) constructed data-derived delineations of places using cell phone call data. Specifically, they mapped the strongest 80% of links among areas within Britain, based on cell phone total talk time. The result is a data-derived division of Britain into social-geographic 'places' at a regional scale.

Advances in heterogeneous network mining have the potential to move beyond simple mapping or network statistics applied to interconnections among places derived from single data sets such as the cell phone data discussed above. Heterogeneous network mining represents multi-typed data as heterogeneous information networks and applies methods that can mine useful knowledge from these networks (Sun and Han 2012).

As one example of the potential for developing rich place-relevant information using this approach, Savelyev and MacEachren (2014, in preparation) have implemented a linked data structure and query mechanism in SensePlace3 (a follow-on to the system discussed above). The implementation supports complex queries across feature types extracted from Twitter data. In Fig. 1, the query for tweets containing the term 'refugee' has been filtered on the basis of a linked query for tweets by individuals who have a profile location of London and that mention Syria in the text of the tweet. The results provide a mix of perspectives from individuals who live in or associate with London. The links on the maps signify all connections among locations mentioned in any of the tweets by individuals from London or any of the tweets mentioning Syria. The results show Ukraine and Iran as other locations of concern across this collective of tweets. While this example focuses only on the data and metadata contained in tweets, the general method of heterogeneous network mining can be applied to explore any forms of connection across data of multiple forms (Janowicz et al. 2012).

Conclusions

Big data, while a potential resource that might enable new understanding of place and interconnections among places has the potential to be used for a wide variety of applications, not all of which will be viewed positively by everyone. Cresswell

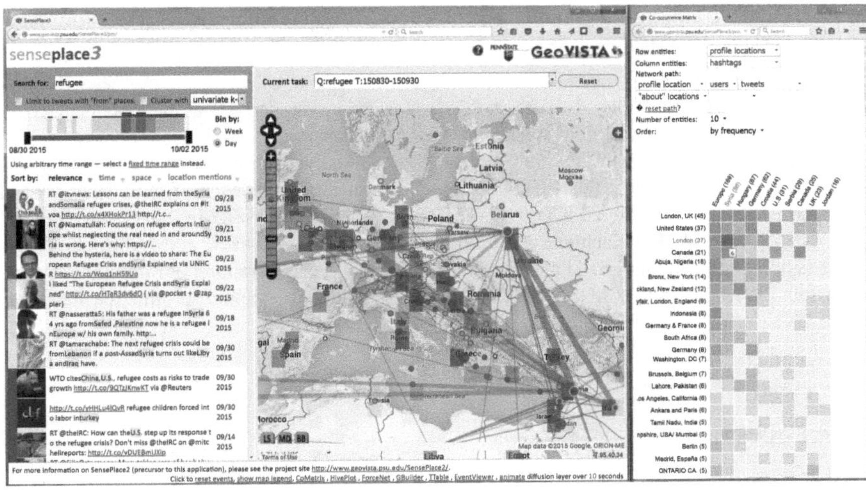

Fig. 1 SensePlace 3 results for tweets that mention "refugee", with a heterogeneous network constraint that identifies the subset of these tweets by individuals with London as their profile location and "Syria" as a term in the tweet text

(2014), for example, quotes Sui and Goodchild (2011) on the start of attempts in GIScience to formalize place in ways that can support application of GIScience methods and tools. He takes a rather critical view, highlighting the ways in which this formalization may be used "in sometimes sinister ways" that include politicians targeting swing voters, supermarkets interrogating shopping habits, and police/security forces sifting personal information in the hope of finding links between crime and place.

It is, of course, important for those of us who develop big data analytical methods to consider the cons as well as the pros of the tools we create. That puts privacy-preserving analytics at the top of the list for important research initiatives as we work to shift the attention of GIScience away from a space-only perspective to one that includes attention to place and the context within which human (and other) activity occurs.

If we can develop methods that minimize the dangers of big data while leveraging the potential, there is an opportunity to address a wide array of place-based challenges for science and society that were impractical to consider prior to the advent of place-aware big data.

I end with two suggested research challenges at the interface of big data, gVA, and place. Research is needed: (1) to integrate advances in methods and technologies that address dynamic, heterogeneous (unstructured + structured), and massively interconnected data to understand place and connections among places at multiple scales; and (2) to create a science of "placial analytics"[1] that addresses place as deeply as GIScience has addressed space thus far.

Acknowledgements Examples in Section "Data Variety" were derived as part of the GeoTxt project, in collaboration with Jan Wallgrün, Morteza Karimzadeh, and Scott Pezanowski; A portion of this research was supported by the Visual Analytics for Command, Control, and Interoperability Environments (VACCINE) project, a center of excellence of the Department of Homeland Security, under Award #2009-ST-061-CI0001. See also: Wallgrün, Jan Oliver, Morteza Karimzadeh, Alan M. MacEachren, Frank Hardisty, Scott Pezanowski, and Yiting Ju. 2014. "Construction and First Analysis of a Corpus for the Evaluation and Training of Microblog/Twitter Geoparsers." GIR'14: 8th ACM SIGSPATIAL Workshop on Geographic Information Retrieval, Dallas, TX.

References

Agarwal, P. (2005). Operationalising "sense of place" as a cognitive operator for semantics in placebBased ontologies. In A. G. Cohn & D. M. Mark (Eds.), *Lecture notes in computer science* (pp. 96–114).

Agnew, J. (2011). *Space and place*. London: The Sage Handbook of Geographical Knowledge (pp. 316–330).

Agnew, J. A., & Duncan, J. S. (2014). *The power of place (RLE Social & Cultural Geography): Bringing together geographical and sociological imaginations*. Routledge.

Alexander, C. (2002). *The nature of order an essay on the art of building and the nature of the universe: Book I-the phenomenon of life* (p. 119). Berkeley, California: The Center for Environmental Structure.

Andrienko, G., Andrienko, N., Bak, P., Keim, D., & Wrobel, S. (2013). *Visual analytics of movement*. Springer Science & Business Media.

Andrienko, G., Andrienko, N., Keim, D., MacEachren, A. M., & Wrobel, S. (2011). Challenging problems of geospatial visual analytics. *Journal of Visual Languages and Computing, 22*(4), 251–256.

Andrienko, N., Andrienko, G., Fuchs, G., & Jankowski, P. (2015). Scalable and privacy-respectful interactive discovery of place semantics from human mobility traces. *Information Visualization, 15*(2), 117–153.

Andriole, S. (2015). Unstructured data: The other side of analytics. *Forbes*. http://www.forbes.com/sites/steveandriole/2015/03/05/the-other-side-of-analytics/print/

Anselin, L. (1995). Local indicators of spatial association—LISA. *Geographical Analysis, 27*, 93–115.

Bennett, B., & Agarwal, P. (2007). Semantic categories underlying the meaning of 'place'. In *International Conference on Spatial Information Theory* (pp. 78–95). New York: Springer.

Bharti, N., Tatem, A. J., Ferrari, M. J., Grais, R. F., Djibo, A., & Grenfell, B. T. (2011). Explaining seasonal fluctuations of measles in Niger using nighttime lights imagery. *Science, 334*(6061), 1424–1427.

[1]placial rather than platial since place is from the old French place and medieval Latin placea (place, spot)—source: http://www.etymonline.com (earlier Latin used platea (courtyard, open space; broad way, avenue) and Greek used plateia (broad way); for comparison, spatial is from the Latin spatium + al (room, area, distance, stretch of time + of or relating to)); analytics rather than analysis since the latter is the activity while analytics, from the Ancient Greek ἀναλυτικά (ànalytiká, is "science of analysis")—source: https://en.wiktionary.org.

Carmona, M. (2003). *Public places, urban spaces: The dimensions of urban design*. Amsterdam: Routledge.

Cervone, G., Sava, E., Huang, Q., Schnebele, E., Harrison, J., & Waters, N. (2016). Using Twitter for tasking remote-sensing data collection and damage assessment: 2013 Boulder flood case study. *International Journal of Remote Sensing, 37*(1), 100–124.

Charlton, M., Fotheringham, S., & Brunsdon, C. (2006). *Geographically weighted regression. NCRM/006, National University of Ireland Maynooth*, Maynooth, Co. Kildare, IRELAND.

Cresswell, T. (2014). *Place: An introduction*. Chichester, West Sussex: Wiley.

De Blij, H. (2008). *The power of place: Geography, destiny, and globalization's rough landscape*. Oxford, UK: Oxford University Press.

Duncan, J. S., & Ley, D. (1993). *Place/culture/representation*. New York, NY: Routledge.

Edwardes, A. J., & Purves, R. S. (2007). Eliciting concepts of *place* for text-based image retrieval. In *Proceedings of the 4th ACM Workshop on Geographical Information Retrieval* (pp. 15–18). ACM, Lisbon, Portugal.

Egenhofer, M., & Herring, J. R. (1990). A mathematical framework for the definition of topological relationships. In *Proceedings of the 4th International Symposium on Spatial Data Handling* (vol. 2, pp. 803–813).

Ellard, C. (2015). *Places of the heart: The psychogeography of everyday life*. New York: Bellevue Literary Press.

Goodchild, M. F. (2011). Formalizing place in geographic information systems. In L. M. M. Burton, S. A. P. Matthews, M. Leung, S. P. A. Kemp, & D. T. T. Takeuchi (Eds.), *Communities, neighborhoods, and health. Social disparities in health and health care* (pp. 21–33). New York: Springer.

Gustafson, P. E. R. (2001). Meanings of place: Everyday experience and theoretical conceptualizations. *Journal of Environmental Psychology, 21*(1), 5–16.

Hardy, D., Frew, J., & Goodchild, M. F. (2012). Volunteered geographic information production as a spatial process. *International Journal of Geographical Information Science, 26*(7), 1191–1212.

Hubbard, P., & Kitchin, R. (2010). *Key thinkers on space and place*. Beverly Hills: Sage.

Hubert, L. J., Golledge, R. G., & Costanzo, C. M. (1981). Generalized procedures for evaluating spatial autocorrelation. *Geographical Analysis, 13*(3), 224–333.

Janowicz, K., Scheider, S., Pehle, T., & Hart, G. (2012). Geospatial semantics and linked spatiotemporal data–past, present, and future. *Semantic Web, 3*(4), 321–332.

Jenkins, A., Croitoru, A., Crooks, A. T., & Stefanidis, A. (2016). Crowdsourcing a collective sense of place. *PLoS ONE, 11*(4), e0152932.

Keim, D., Kohlhammer, J., Ellis, G., & Mansmann, F. (Eds.). (2010). *Mastering the information age: Solving problems with visual analytics*. Goslar, Germany: Eurographics Association.

Keim, D. A., Krstajic, M., Rohrdantz, C., & Schreck, T. (2013). Real-time visual analytics for text tsreams. *Computer, 46*(7), 47–55.

Klippel, A., Yang, J., Wallgrün, J.O., Dylla, F., & Li, R. (2012). *Assessing similarities of qualitative spatio-temporal relations. Spatial Cognition VIII* (pp. 242–261). New York: Springer.

Kohlhammer, J., May, T., & Hoffmann, M. (2009). Visual analytics for the strategic decision making process. *GeoSpatial Visual Analytics*, 299–310.

Kraft, T., Wang, D. X, Delawder, J., Dou, W., Yu, L., & Ribarsky, W. (2013). Less after-the-fact: Investigative visual analysis of events from streaming twitter. In B. Geveci, H. Pfister & V. Vishwanath (Eds.), *IEEE symposium on large-scale data analysis and visualization (LDAV)* (pp. 95–103), Citeseer.

Langran, G. (1992). *Time in geographic information systems. Technical topics in geographic information systems*. London: Taylor & Francis.

Liu, Y., Liu, X., Gao, S., Gong, L., Kang, C., Zhi, Y., et al. (2015). Social sensing: A new approach to understanding our socioeconomic environments. *Annals of the Association of American Geographers, 105*(3), 512–530.

Madia, K. (2015). *Embracing real-time, streaming analytics in the insight economy, blogs*. IBM, IBM Big Data & Analytics Hub.

Malik, A., Maciejewski, R., Jang, Y., Oliveros, S., Yang, Y., Maule, B., et al. (2012). A visual analytics process for maritime response, resource allocation and risk assessment. *Information Visualization, 13*(2), 93–110.

Martin, D. (1998). Optimizing census geography: The separation of collection and output geographies. *International Journal of Geographical Information Science, 12*(7), 673–685.

Massey, D. (2013). *Space, place and gender*. New York: Wiley.

Massey, D. B. (1994). *Space, place, and gender*. Minneapolis: University of Minnesota Press.

Mennis, J. L., Peuquet, D. J., & Qian, L. J. (2000). A conceptual framework for incorporating cognitive principles into geographical database representation. *International Journal of Geographical Information Science, 14*(6), 501–520.

Monroe, B. L. (2013). The five Vs of big data political science introduction to the virtual issue on big data in political science. *Political Analysis, Virtual Issue, 19*(5), 66–86.

Morison, B. (2002). *On location: Aristotle's concept of place*. Oxford Aristotle Studies: Clarenden Press, Oxford, Oxford. 194.

Nairn, K., Kraftl, P., & Skelton, T. (eds.) (2016). *Space, place and environment. Geographies of children and young people* (vol. 3). New York: Springer.

Nelson, J. K., Quinn, S., Swedberg, B., Chu, W., & MacEachren, A. M. (2015). Geovisual analytics approach to exploring public political discourse in Twitter. *ISPRS International Journal of Geo-Information, 4*(1), 337–366.

Peuquet, D. J. (1988). Representations of geographic space: Toward a conceptual synthesis. *Annals of the Association of American Geographers, 78*(3), 373–394.

Pezanowski, S., MacEachren, A. M., Savelyev, A., & Robinson, A. C. submitted. *SensePlace3: A geovisual framework to analyze place-time-attribute information in social media*.

Pred, A. (1984). Place as historically contingent process: Structuration and the time- geography of becoming places. *Annals of the AAG, 74*(2), 279–297.

Quinn, S., & Yapa, L. (2015). OpenStreetMap and food security: A case study in the city of Philadelphia. *The Professional Geographer, 68*(2), 271–280.

Ratti, C., Sobolevsky, S., Calabrese, F., Andris, C., Reades, J., Martino, M., et al. (2010). Redrawing the map of Great Britain from a network of human interactions. *PLoS ONE, 5*(12), e14248.

Relph, E. (1976). *Place and placelessness, 1*. Pion.

Richardson, M., Kamalski, J., Huggett, S., & Andrew, P.... Courtesy of In, e.b. (2012). The fundamental interconnectedness of all things. In K. Börner & M. J. Stamper (Eds.), *Courtesy of Elsevier Ltd. In "8th iteration (2012): science maps for kids," places & spaces: Mapping science. curated by the cyberinfrastructure for network science center*, http://scimaps.org

Robinson, A. C. (2011). Supporting synthesis in geovisualization. *International Journal of Geographical Information Science, 25*(2), 211–227.

Roche, S., & Rajabifard, A. (2012). Sensing places' life to make city smarter. In *Proceedings of the ACM SIGKDD International Workshop on Urban Computing* (pp. 41–46). Beijing, China: ACM.

Savelyev, A., & MacEachren, A. M. (2014). Interactive, browser-based information foraging in heterogeneous space-centric networks. In G. Andrienko et al. (Eds.), *Workshop on geovisual analytics: Interactivity, dynamics, and scale, in conjunction with GIScience 2014*. Vienna: Austria.

Savelyev, A., & MacEachren, A. M. in preparation. Interactive, browser-based information foraging in heterogeneous space-centric networks.

Scheider, S., & Janowicz, K. (2014). Place reference systems. *Applied Ontology, 9*(2), 97–127.

Sui, D., & Goodchild, M. (2011). The convergence of GIS and social media: Challenges for GIScience. *International Journal of Geographical Information Science, 25*(11), 1737–1748.

Sun, Y., & Han, J. (2012). Mining heterogeneous information networks: Principles and methodologies. *Synthesis Lectures on Data Mining and Knowledge Discovery, 3*(2), 1–159.

Thomas, J. J., & Cook, K. A. (Eds.). (2005). *Illuminating the path: The research and development agenda for visual analytics*. Los Alamos, CA: IEEE Computer Society.

Tomaszewski, B., Blanford, J., Ross, K., Pezanowski, S., & MacEachren, A. M. (2011). Supporting geographically-aware web document foraging and sensemaking. *Computers, Environment and Urban Systems, 35*(3), 192–207.

Tomaszewski, B., & MacEachren, A. M. (2012). Geovisual analytics to support crisis management: Information foraging for geo-historical context. In *Information Visualization (invited extension of paper originally published in proceedings of IEEE VAST 2010)* (vol. 11 (4), pp. 339–359).

Tuan, Y.-F. (1977). *Space and place: The perspective of experience*. Minneapolis: University of Minnesota Press. 226.

Vasardani, M., & Winter, S. (2016). Place properties. In H. Onsrud & W. Kuhn (Eds.), *Advancing geographic information science: The past and next twenty years* (pp. 243–254). Needham, MA: GSDI Association Press.

Wang, X., Miller, E., Smarick, K., Ribarsky, W., & Chang, R. (2008). Investigative visual analysis of global terrorism. *Computer Graphics Forum, 27*(3), 919–926.

Winter, S., & Freksa, C. (2012). Approaching the notion of place by contrast. *Journal of Spatial Information Science, 2012*(5), 31–50.

Wojcik, S. P., Hovasapian, A., Graham, J., Motyl, M., & Ditto, P. H. (2015). Conservatives report, but liberals display, greater happiness. *Science, 347*(6227), 1243–1246.

Woodcock, C. E., & Strahler, A. H. (1987). The factor of scale in remote sensing. *Remote Sensing of Environment, 21*(3), 311–332.

Xu, Y., Shaw, S.-L., Zhao, Z., Yin, L., Lu, F., Chen, J., et al. (2016). Another tale of two cities: Understanding human activity space using actively tracked cellphone location data. *Annals of the American Association of Geographers, 106*(2), 489–502.

Yang, X., Ye, X., & Sui, D. Z. (2016). We know where you are: In space and place-enriching the geographical context through social media. *International Journal of Applied Geospatial Research (IJAGR), 7*(2), 61–75.

A Comparative Study of Various Properties to Measure the Road Hierarchy in Road Networks

Xun Wu, Hong Zhang, Yunhui Xu and Jie Yang

Introduction

Spatial data can be represented at different scales, which may facilitate map navigation and spatial analysis. Fully automated transformation of a map from one scale to a smaller scale is still a research topic of interest in the field of mapping and cartography (Li 2006). This study is concerned with selective omission in road network data, because road is one of the most important geographical features on a map, and selective omission (meaning the retention of more important roads) is an operation necessary for automated road network generalization.

Selective omission in a road network has been the subject of extensive studies. Some researchers analyzed road segments (Mackaness and Beard 1993; Mackaness 1995; Thomson and Richardson 1995) or road intersections (Mackaness and Machechnie 1999) for selection, because a road network is always stored in a database as intersections and segments. Some workers built strokes, which are defined as 'a set of one or more arcs in a non-branching and connected chain' (Thomson and Richardson 1999), and the selections were based on those strokes. The use of strokes makes possible the analysis of road networks based on the importance of individual roads, even in the absence of all other thematic information (Thomson and Brooks 2007). The importance of each stroke may be determined by various properties, such as road length, stroke connectivity (Zhang 2004a), degree, closeness, and betweenness centralities (Jiang and Harrie 2004). Now most researchers propose integrated indicators with various road properties.

X. Wu (✉) · H. Zhang · Y. Xu · J. Yang
Faculty of Geosciences and Environmental Engineering, Southwest Jiaotong University, Chengdu 611756, People's Republic of China
e-mail: xunwu0123@163.com

C. Zhou et al. (eds.), *Spatial Data Handling in Big Data Era*,
Advances in Geographic Information Science,
DOI 10.1007/978-981-10-4424-3_11

A method based on complex network analysis was proposed to estimate the hierarchies of urban road networks, in which the degree, closeness, betweenness and length are considered (Luan et al. 2012; Liu et al. 2014; He et al. 2015). An integrated approach was proposed in which different structures or patterns in a road network are considered (Li and Zhou 2012; Yang et al. 2013). Considering the connectivity and the geometric structure of the road network, the functionality of the stroke was the basis of road selection (Xu et al. 2012).

However, it is short of evaluations on the situation (aspects of network functionality and cartography) in which the new composite indexes fit. To our knowledge, no literature has focused on a comparative analysis on the composite indexes and finding a new composite index to define the importance of a road in view of network structure functionality, which is the main concern of this study.

The study is organized as follows. In Section "Linear Correlation Model of Road Ranking", a brief description of the road ranking approaches, evaluations and study area is provided. Section "Experiment Result and Analysis" shows the experiment results. Finally, conclusions are drawn and some future work is given.

Methodology and Experiment Design

In order to achieve the comprehensive evaluation on the measurement of the road importance of a real road network, one typical road ranking approach and a new measure are used for the evaluation. In this section, the approach, measure and study area are briefly introduced.

The stroke-based approach was first proposed by Thomson and Richardson (1999). This approach has two steps, building the strokes and ordering the strokes. Building the strokes means concatenating continuous and smooth road segments into a whole. Ordering the strokes means ranking the strokes in a descending order from high to low importance. It is crucial to evaluate the importance of the stroke.

Dual Graph of the Road Network

In recent years, complex networks have been gradually applied to transportation and GIS, contributing to a deep analysis on the complexity and functionality of the structure of road network. Figure 1 shows different network topology structures of the same road network. Compared to the generated topology of the network (Fig. 1b), the dual graph (Fig. 1c) could be used to analyze the network structure and functionality further (Boccaletti et al. 2015). And the dual graph has an advantage of analyzing the connectivity and reliability of a road network and the importance of the road in the real road network.

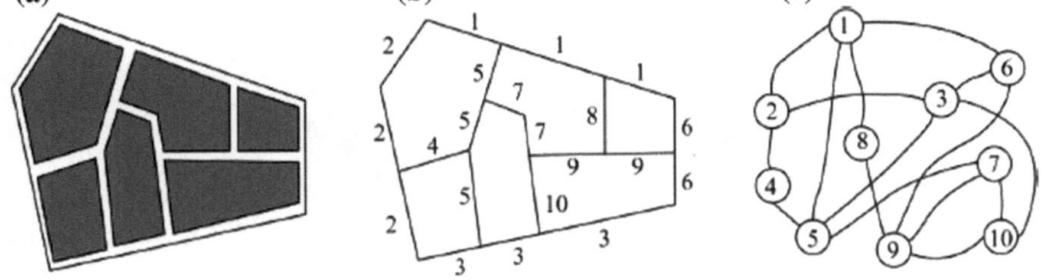

Fig. 1 Transportation network topology structure. **a** is the real road network; **b** is the generated topology of the network and; **c** is the dual graph of the network

Fig. 2 Example of centrality maximums

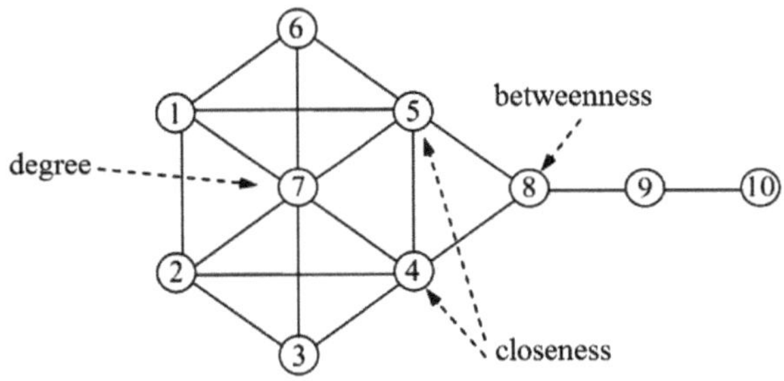

Structural and Geometric Characteristics of the Road Network

Usually the network centrality is used as the index to analyze the structure characteristics of complex network. There are three basic indexes of centrality: degree, betweenness and closeness, as shown in Fig. 2. The clustering coefficient is an important index, which is also considered in this paper. So, four structure indexes and one geometric index are used to evaluate the importance of the stroke.

(1) Centrality of degree is expressed as follows:

$$Degree = D_i = \sum_{j=1}^{n} \delta_{ij} \tag{1}$$

where, δ_{ij} shows whether stroke i intersects with stroke j. If they intersect, δ_{ij} is 1, otherwise 0. In the structure analysis on the road network, the greater the value of degree is, the more the road connections are. The degree plays a significant role in the entire road network.

(2) Centrality of closeness is expressed as follows:

$$Closeness = C_i = 1/\sum_{j=1, j\neq i}^{n} n_{ij} \tag{2}$$

where, n_{ij} is the number of the strokes included in the shortest path from i to j. The closeness is a global measurement that indicates the center of a city. High-rank roads should exhibit good accessibility to other roads. Compared with the centrality of degree, the closeness could further describe the accessibility of a stroke to its indirectly connected strokes. The greater the index value is, the more extensive range of services and the impacts of the stroke are, and the rank of the stroke is higher.

(3) Centrality of betweenness is expressed as follows:

$$Betweenness = B_i = 1/\sum_{j\neq k\neq i}^{n} n_{jk}(i)/n_{jk} \tag{3}$$

where, n_{jk} is the number of the strokes included in the shortest path from j to k, and $n_{jk}(i)$ is the number strokes in the shortest path (i to j) passing the i. In the road network, the stronger the betweenness of the stroke is, representing more passing times on the shortest path, the more obvious influences like bridges and hubs are.

(4) Clustering coefficient is expressed as follows:

$$CC_i = 2e_i/k_i(k_i - 1) \tag{4}$$

where, k_i is the degree of the stroke, and e_i is the number of triangles formed between any two neighbors. Different from the centrality of degree, the smaller the clustering coefficient is, the greater the functional role the node plays is in the network.

Length of the stroke is constructed by length of the continuous and smooth road segments, which is the geometric property. The longer length of the stroke, the higher the rank in the road networks.

Linear Correlation Model of Road Ranking

The five properties can only reflect some aspects of ranking of road networks in terms of structural and geometric characteristics. A series of correlation models of road ranking with structural and geometric characteristics have been built and could be used to comprehensively assess the ranking of road networks. This paper only utilizes a simple and basic linear correlation model of road ranking, which is expressed as:

$$Rank = \sum_{i=1}^{n} \alpha_i X_i \tag{5}$$

where, X_i is the properties of the stroke, and α_i is the weight factor of each property. Five properties of the stroke are used to rank the stroke. One is the basic geometric property, and the others are structural properties. The thematic property, which is unavailable, is not considered in the stroke-based approach. In order to keep a principle that the amount of information of the model could be maximized (Luan 2012), α_i is defined as follows:

$$\alpha_i = E_i / \sum_{j=1}^{m} E_i \tag{6}$$

The information of the property can be obtained from E_i, which is expressed as follows:

$$E_i = \sigma_i \sum_{j=1}^{m} (1 - r_{ij}) \tag{7}$$

The standard deviation of the property is defined as σ_i. The r_{ij} is the correlation coefficient between the properties.

Another way can explain α_i is shown in Eq. 8. In the expression, μ_i is the mean of each property, but the coefficient of variation is not considered.

$$\alpha_i = \sigma_i / \mu_i \tag{8}$$

Measurement

Using the linear correlation model of road ranking, geometric and structure properties could be chosen to integrate a new index of road ranking based on the stroke. However, it is not the best solution to choose all properties. Therefore, we should select some properties and combine them into a new index. Also, a method is needed to evaluate the index.

In complex networks, there are several ways of measuring the functionality of the networks. One key quantity is the average inverse geodesic length (Holme et al. 2004), which is a finite value even for a disconnected graph:

$$l^{-1} = \frac{1}{N(N-1)} \sum_{v \not\subset V} \sum_{w \neq v \in V} \frac{1}{d(v,w)'} \tag{9}$$

The road network can be expressed as a graph: $g = (v,e)$, where v is the set of the vertices that stands the roads. Each edge connects exactly one pair of vertices and represents the connection relation of each road. The $d(v,w)$ is the length of the geodesic between v and w. When we remove the high-rank roads, the functionality of the network could go downhill in a quick manner and then in a slower pace. So, the l^{-1} can be used to measure the indexes by removing the high-rank roads orderly.

Study Area

Three real road networks of varying patterns are tested (Fig. 3). After building the strokes, the road network of Chengdu (Fig. 3a) has 253 strokes, Hong Kong 484, and New York 933.

Evaluations on Road Ranking Using Road Removing

Steps of Road Removing

The detailed description of the evaluations on road ranking using the road removing is as follows:

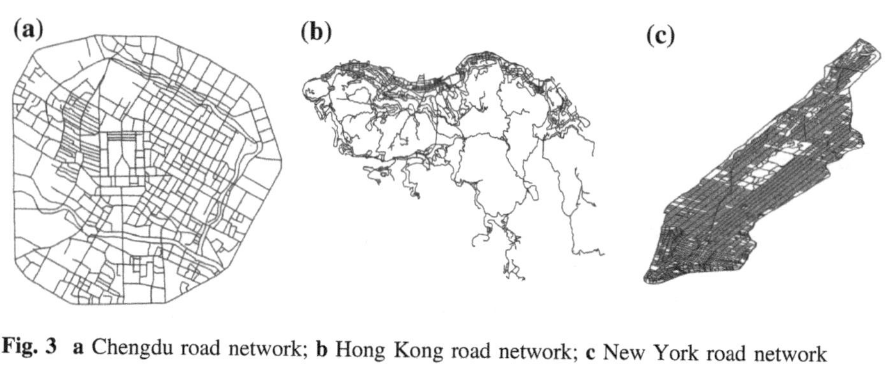

Fig. 3 **a** Chengdu road network; **b** Hong Kong road network; **c** New York road network

(1) Building the strokes: Set the threshold of angle as 45 degrees and build the strokes for the road network. These are the basic operations of studying the road network.
(2) Building the dual graph: Adopt the dual method to build the dual graph of the road network based on the strokes.
(3) Calculating the values of properties: Obtain the values of degree, betweenness, closeness, clustering coefficient and the length of the stroke (the node on the dual graph).
(4) Generating the integrated indexes of road ranking: Apply the properties to the linear correlation model to generate eight integrated indexes that may be used to rank the roads and calculate the indexes respectively. Here, the length and the degree of the stroke as the basic elements of road ranking should be considered. Table 1 lists the eight integrated indexes that adopt different properties.
(5) Removing the strokes in order: Sort the strokes by values of the integrated indexes respectively in descending order and calculate the l^{-1} by removing the stroke in descending order respectively. Plot the change curve of the l^{-1}.

Experiment Result and Analysis

Eight indexes are utilized and three real road networks of different patterns are tested as shown in Fig. 4. Figure 4 shows the change curve of l^{-1} when the roads are removed in descending order. That is, the functionality of the real road network could be reflected by the curve. In order to give a clear observation, the result of each road network are represented by two graphs, where all index tests are included.

The results of the Chengdu road network (Fig. 4a, b) show that if clustering coefficient is added into the composite index, the l^{-1} does not go downhill when the roads of Ranks 20–30 are removed; while the DL and DLB perform very well. In

Table 1 Eight different indexes using five properties of the road

Index	Length	Degree	Betweenness	Closeness	Clustering coefficient
DL	+	+			
DLB	+	+	+		
DLCC	+	+			+
DLBCC	+	+	+		+
DLC	+	+		+	
DLCB	+	+	+	+	
DLCCC	+	+		+	+
DLCBCC	+	+	+	+	+

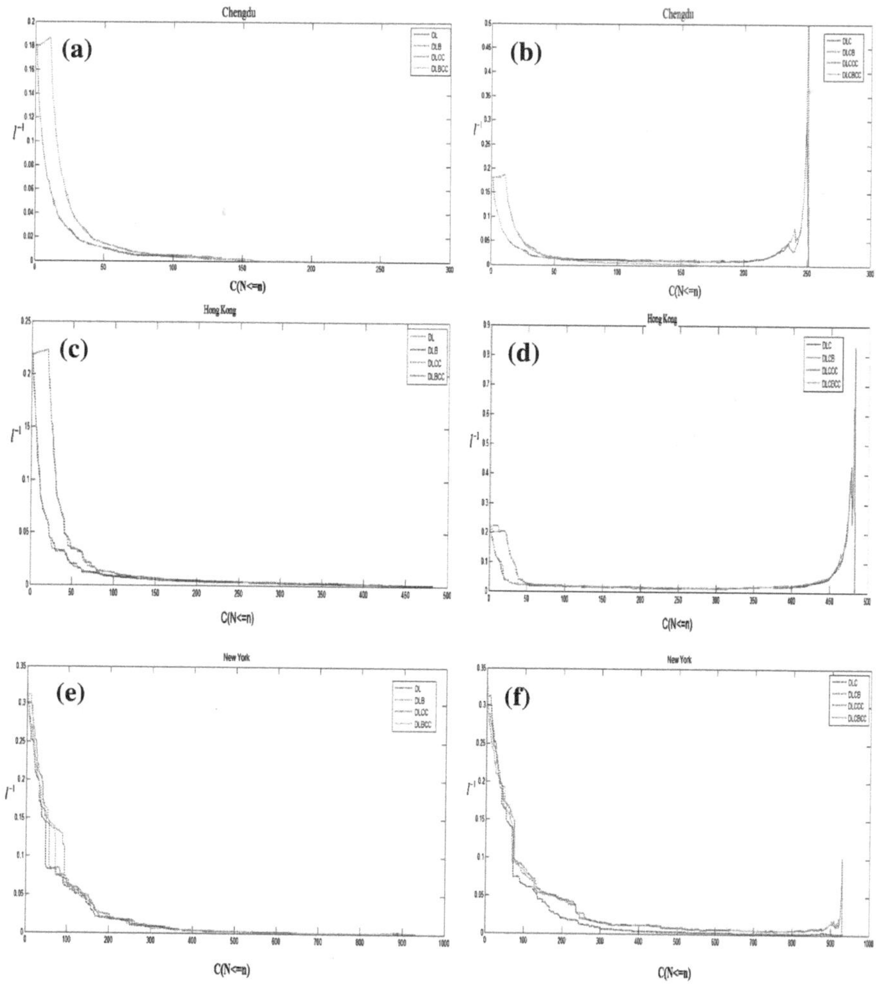

Fig. 4 Eight indexes are tested by the l^{-1} for three different road patterns. **a** and **b** are for Chen Du; **c** and **d** for Hong Kong; **e** and **f** for New York

addition, if the closeness is considered, the l^{-1} exhibits a jump after low-rank roads are removed. As shown in Fig. 4c and d, the similar phenomena occur in the Hong Kong road network. Since there are more roads in New York, an illusion may be given to us that New York behaves dissimilarly with Chengdu. However, if you zoom out Fig. 4e and f, you can find the same phenomenon.

To further test the validity of this indicator (DLB), a road selection test is carried out for the Chengdu road network in different selection proportions. Figure 5 gives five results of road selection. We can see that: (1) In each proportion, even a very small proportion, the selected network could maintain the topology connectivity of the original network and cover the whole range of the original road network;

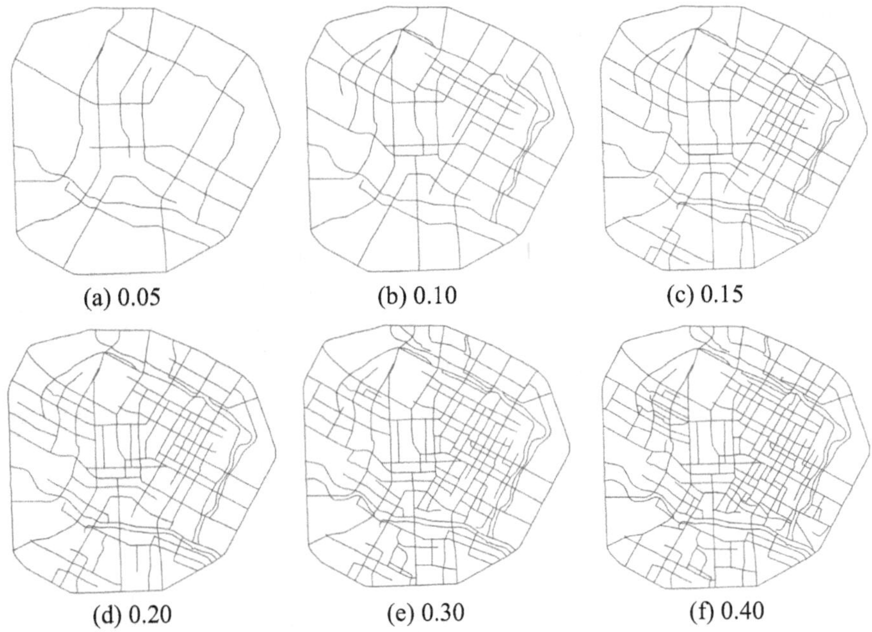

Fig. 5 Road selection results at various selection ratios

(2) In each proportion, the selected road network could keep the overall structure of the original road network. (3) As the selected proportion increases, the added roads are more reasonable with the density and the overall structure of the original road network being considered. And the hierarchy of the road network may be reflected. The Hong Kong and New York road networks involve the same phenomenon.

Conclusions

Evaluating the road rank is not simply aggregating many properties of roads. We should also consider whether some properties need to be added into composite indexes. The result shows that the length and degree are the basis for evaluating the importance of roads. If the clustering coefficient is considered, composite indexes have adverse effects on the sorting of high-rank roads. While the closeness is added, the sorting of low-rank road is unreasonable. If the length, degree and betweenness are considered all together, the composite indexes perform best in the sorting of roads.

Furthermore, in order to enhance the performance of the road ranking method, it is of great value to take more road ranking approaches (not only the linear correlation model) and more characteristics of the road networks into account.

Acknowledgements This work is jointly supported by the National Natural Science Foundation of China project (No.41101361 and 41471383) and the Fundamental Research Funds for the Central Universities (No.SWJTU11CX063).

References

Boccaletti, S., Latora, V., Moreno, Y., et al. (2015). Complex networks: Structure and dynamics. *Physics Reports, 424*(4–5), 175–308.

Holme, P., Kim, B. J., Yoon, C. N., et al. (2004). Attack vulnerability of complex networks. *Physical Review E Statistical Nonlinear & Soft Matter Physics, 65*(5), 634–634.

He, H. W., et al. (2015). Road selection based on road hierarchical structure control. *Acta Geodaetica et Cartographica Sinaica, 44*(4), 453–461.

Jiang, B., & Harrie, L. (2004). Selection of roads from a network using self-organizing maps. *Transactions in GIS, 8*(3), 335–350.

Li, Z. L. (2006). *Algorithmic foundation of multi-scale spatial representation* (p. 280). Raton: CRC Press (Taylor & Francis Group).

Li, Z. L., & Zhou, Q. (2012). Integration of linear and areal hierarchies for continuous multi-scale representation of road networks. *International Journal of Geographical Information Science, 26*(5), 855–880.

Liu, G., et al. (2014). Auto-selection method of road network based on evaluation of node importance for dual graph. *Acta Geodaetica et Cartographica Sinaica, 43*(1), 97–104.

Luan, X., Yang, B., & Zhang, Y. (2012). Structural hierarchy analysis of streets based on complex network theory. *Geomatics & Information Science of Wuhan University, 37*(6), 728–732.

Mackaness, W. (1995). Analysis of urban road networks to support cartographic generalization. *Cartography and Geographic Information Science, 22*(4), 306–316.

Mackaness, W. A., & Beard, K. M. (1993). Use of graph theory to support map generalization. *Cartography and Geographic Information Science, 20*(4), 210–221.

Mackaness, W., & Mackechnie, G. (1999). Automating the detection and simplification of junctions in road networks. *GeoInformatica, 3*(2), 185–200.

Thomson, R., & Brooks, R. (2007). Generalisation of geographical networks. In A. Ruas & W. A. Mackaness & L. T. Sarjakoski (Eds.), *Chapter 13 in generalization of geographic information: Cartographic modeling and applications* (pp. 255–267). Amsterdam: Elsevier.

Thomson, R., & Richardson, D. (1999). The "good continuation" principle of perceptual organization applied to the generalization of road networks. In *Proceedings of the 19th International Cartographic Conference. Ottawa* (pp. 1215–1223), 14–21 August 1999.

Thomson, R., & Richardson, D. (1995). A graph theory approach to road network generalisation. In: *Proceeding of the 17th International Cartographic Conference. Barcelona* (pp. 1871–1880), 3–9 September 1995.

Yang, M., et al. (2013). A method of road network generalization considering stroke properties of road object. *Acta Geodaetica et Cartographica Sinaica, 42*(4), 581–587.

Xu, Z., Liu, C., Zhang, H., et al. (2012). Road selection based on evaluation of stroke network functionality [J]. *Acta Geodaetica et Cartographica Sinica, 41*(5), 769–776.

Zhang, Q. (2004a). Road network generalization based on connection analysis. In *The 11th International Symposium on Spatial Data Handling. Leicester* (pp. 343–353), 23–25 August 2004.

Detail Resolution: A New Model to Describe Level of Detail Information of Vector Line Data

Xiaoqiang Cheng, Huayi Wu, Tinghua Ai and Min Yang

Introduction

Scale is an important factor in almost all kinds of scientific research questions. In respect of geospatial information, map scale is the most common word representing scale. Map scale which is expressed by numerator divided by denominator and commonly used in traditional cartography and vector-based GIS applications, denotes the ratio of map distance to real world distance. Meanwhile, this scale representation is used to describe the level of detail (*LoD*) information of vector dataset, on the condition that a dataset's source and processing flow are explicitly known. However, scale in this form is incompetent as the proliferation of volunteered geographic information (VGI) and web-based services.

GIS data created by volunteers is criticized by low quality issues, such as incompleteness (Hecht et al. 2013), inconsistency, heterogeneity (Touya and Brando-Escobar 2013) and missing of metadata, etc. What's the map scale of a road curve uploaded by user? Do the roads within one road network share a same map scale? Is the river feature's map scale same with the road feature? These questions

X. Cheng (✉)
Hubei University, Wuhan 430062, People's Republic of China
e-mail: carto@whu.edu.cn

H. Wu · T. Ai · M. Yang
Wuhan University, Wuhan 430079, People's Republic of China
e-mail: wuhuayi@whu.edu.cn

T. Ai
e-mail: tinghua_ai@163.net

M. Yang
e-mail: 250268582@qq.com

C. Zhou et al. (eds.), *Spatial Data Handling in Big Data Era*,
Advances in Geographic Information Science,
DOI 10.1007/978-981-10-4424-3_12

are still unknown. So that, it is impossible to describe a VGI dataset's *LoD* using map scale as traditional GIS does. As a consequence, VGI data' handling in map generalization and visualization is held back due to the missing of accurate *LoD* information.

Aiming at dealing with the visualization of VGI data, this study proposes a new model supporting *LoD* information detection and representation of vector GIS data, in case of line features. Current digital devices including personal computer, cell phone and tablet are all raster display devices which consist of a matrix of pixels. Visualizing vector data on these devices needs mapping geometric coordinates in real numbers to pixel coordinate in integer. This mapping process called rasterization causes loss of information, which is called "aliasing" in computer graphics. In geospatial visualization, aliasing falls into two categories: jaggy boundaries that can be solved by regular antialiasing techniques (MEI et al. 2008) and coalescence which is usually considered in map generalization (Shea and McMaster 1989). Visual coalescence usually manifests that geographic features or feature's parts are too dense to discern. Coalescences are scale-dependent, that is, when changing the scale by zooming in, the number of coalescences decreases, until all coalescences fade away. So we make an assumption that the resolution at which a feature's coalescence just disappears should be used to represent the *LoD* information of this feature. There were studies using devices' resolution to operate map generalization (Li and Openshaw 1993), however, they were still constrained by map scale.

Methodology

Modern computer graphics systems are based on raster display which consists of matrix of pixels. The size of matrix is called resolution of raster display. Resolution is an important indicator of raster display and is represented by the width and height of pixel matrix, e.g. 1024 × 768. Drawing vector GIS data on raster display needs a classic algorithm called rasterization, which is the task of taking an image described in a vector graphics format (shapes) and converting it into a raster image (pixels or dots) for output on a video display or printer, or for storage in a bitmap file format.[1]

For geographic data, rasterization actually converts geographic coordinates in real numbers to pixel coordinates in integers, so a pixel always corresponds to a certain amount of geographic distances. In digital map visualization, we call the distances represented by one pixel **Map Resolution** (R_{map}). This distance is decided by one important parameter of rasterization algorithm, called *cellsize*, which means how much geographic distance is mapped to one pixel. The smaller the *cellsize*, the raster line approximates the vector line better and the bigger the *cellsize*, the raster line discards more detail and distorts more seriously. Vector line in Fig. 1-I is

[1]https://en.wikipedia.org/wiki/Rasterisation.

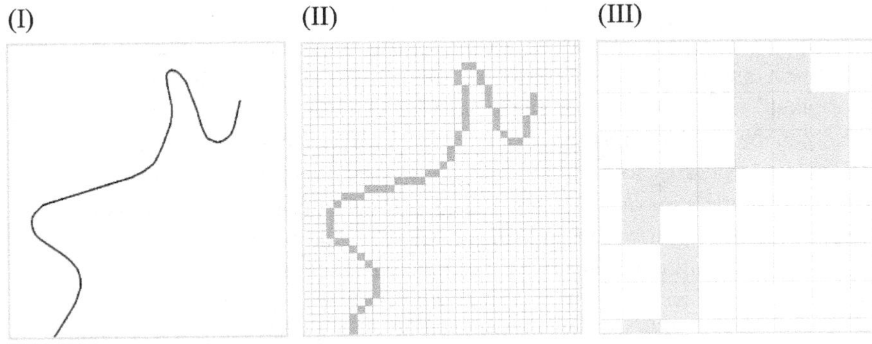

Fig. 1 *Cellsize*'s influence on visualization of vector data

rasterizated to II and III with different *cellsize* respectively. Obviously, raster line in II is more legible than raster line in III.

Loss of detail and distortion in Fig. 1-III were essentially caused by mapping continuous variable to discrete variable during rasterization. In computer graphics, distortions arising from "continuous to discrete" mapping are called aliasing and techniques improving or eliminating these distortions are called anti-aliasing. There are two common aliasing problems in geospatial visualization, one is jagged curve and another is visual congestion or coalescence. This study mainly concerns the latter. As mentioned by Shea (Shea and McMaster 1989), coalescence is a condition where features will touch as a result of either of two factors: the separating distance is smaller than the resolution of the output device or the features will touch as a result of the symbolization process. Coalescence is scale-dependent and that is to say, coalescences only exist across a fixed scale range and disappear after zooming in adequately. So for a geographic feature with coalescent visualization, there must be a scale point at which the coalescence disappears and this scale point can be used to denote the *LoD* information of the geographic feature. To validate the assumption, this paper introduces a new measure called degree of coalescence (DoC) to quantify the coalescence and to detect the scale point automatically.

Degree of Coalescence

This degree of coalescence measure is designed on the basis of rasterization algorithm. The basic idea is as follows: a vector curve doesn't have width, however, a raster line converted from vector curve has width (1 pixel at least), area (the number of all pixels) and boundary. Pixels in a raster line are classified into categories: interior pixels and exterior pixels. If a pixel has 4 adjacent pixels in its 4-neighborhood, it is an interior pixel, otherwise, it is an exterior pixel. For a vector curve conforming to OGC simple feature specification (OGC 2011), when it is converted to a raster line with *cellsize* small enough, the raster line is not coalescent and all pixels should locate on the boundary of the raster line, in brief, all pixels are "exterior pixel". On the contrary, if the *cellsize* is not small enough, coalescence will appear and some pixels will be surrounded by other pixels and become "interior pixel". The ratio between the number of exterior pixels and the number of all pixels can be used to measure the degree of display clarity. A *DoC* calculation formula is defined as formula (1). $N_{boundary}$ is the number of exterior pixels and N_{sum} denotes the sum of all pixels. $N_{boundary}$ is calculated by tracing the one side boundary along the direction and tracing another side backward, so all pixels in a raster line without coalescence will be counted twice, that is, $N_{boundary} = 2 \cdot N_{sum}$. There are pixels counted once or ignored when a raster line is coalescent, so $N_{boundary} < 2 \cdot N_{sum}$. We use the ***bwboundaries*** function in **Matlab** to calculate the $N_{boundary}$. As shown by formula (1), the higher the *DoC*, the raster line is more coalescent; the smaller the *DoC*, the raster line is clearer.

$$DoC = 1 | \frac{N_{boundary}}{2 \cdot N_{sum}} \quad (1)$$

Different *cellsizes* lead to different *DoCs*. Four curves with different complexity are selected to demonstrate the influence of *cellsize*. Curve I, II, III and VI in Fig. 2 are all rasterizated in four *cellsizes* (initial *cellsize*, double, quadruple and octuple). As can be seen from Fig. 2, curve I with low complexity is still legible when *cellsize* is quadruple, while curve VI with high complexity is coalescent at initial *cellsize*. Moreover, curve VI's readability gets lower and *DoC* gets higher as the *cellsize* increases. In sum, the *DoC* reflects people's perception of coalescence correctly and *DoC* is suitable for *LoD* detection and representation.

Detail Resolution

As mentioned above, small *cellsize* creates clear raster line and big *cellsize* creates coalescent raster line. There should be a critical *cellsize* at which the raster line becomes coalescent from clear as the *cellsize* increases. So we first set a threshold of *DoC* to define whether a raster line is clear or not and then adjust the *cellsize* to approach *DoC* to the threshold. The *cellsize* at which the *DoC* equals or approximates the threshold can be used to express the *LoD* information of a vector line. This *LoD* representation method concerns the reservation of detail in visualization and *cellsize* means map resolution, so we define it as Detail Resolution (DR). This paper sets 0.1 as the *DoC* threshold. The calculation algorithm of DR is as follows.

(1) Rasteratize vector geographic feature and get feature's raster from. Rasterization's initial *cellsize* is the lager value of width and height of minimum bounding rectangle dividing feature's vertex number.
(2) Using *bwboundaries* function in Matlab to get the $N_{boundary}$ and to calculate the *DoC*.

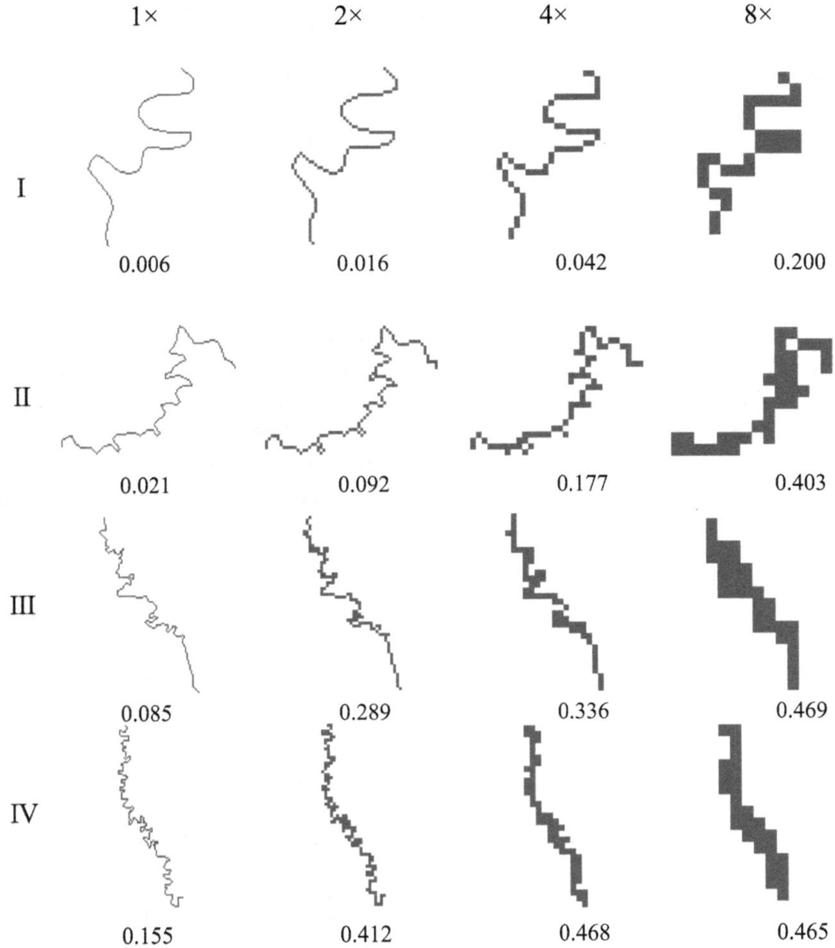

Fig. 2 Cellsize' influence on DoC and corresponding visual effects

(3) If the *DoC* is smaller than the threshold, magnify the *cellsize* 10%; otherwise, minify the *cellsize* 10%. Repeat step (1) and (2). 10% is step size which can be adjusted on demand.
(4) Repeat step (3) until the *DoC* approaches to the threshold. use the *cellsize* at which *DoC* is closest to and smaller than the threshold as the DR of geographic feature.

At this point, three resolution concepts are given. Device Resolution's object is raster display, means the physical size of a single pixel. Map Resolution's object is digital map, denotes the geographic distance of a single pixel. Detail Resolution's object is vector geographic feature, is the largest *cellsize* of rasterization algorithm with legible visualization.

The calculation of DR is independent of specific output raster display or device, however, its application must take parameters of device into consideration. Size and device resolution are two main indices of raster display. Size determines user's visual distance and the minimum discernible unit, device resolution signifies the capability of depicting detail. High device resolution means low probability of visual coalescence. The higher the device resolution, the smaller the physical size of a single pixel and it is too small to identify a pixel at the normal visual distance. For devices with high resolution, drawing a vector feature at its DR can get raster line not coalescent at pixel level, while clarity cannot be guaranteed from user's perception. It is necessary to introduce device's influence on DR. assuming map resolution is $\boldsymbol{R_{map}}$, the target device' minimum discernible distance is $\boldsymbol{d}$ pixels, only features whose DR are bigger than $d{\cdot}\boldsymbol{R_{map}}$ are legible. Map symbolization's influence on DR can be quantified similarly. If line width is w pixels, features whose DR are bigger than $d{\cdot}w{\cdot}\boldsymbol{R_{map}}$ can be visualized clearly. When vector features are drawn on raster display, it is easy to judge the quality of visualization through comparing DR and R_{map}.

Experiments

DR's computation and application are separated in paradigm of DR (Fig. 3). DR's calculation is rasterization algorithm essentially, although it is compute-intensive due to multiple iteration of different *cellize*s, DR's calculation is device-independent. Data collection and update on server-side should trigger the calculation of DR which is stored as an attribute of geographic features to guarantee the timeliness of *LoD* information. DR's application is device-dependent. DR's application is simple comparison of numeric values and suitable for client-side claiming high efficiency. The following sections will give two case studies of DR's calculation and application in detail.

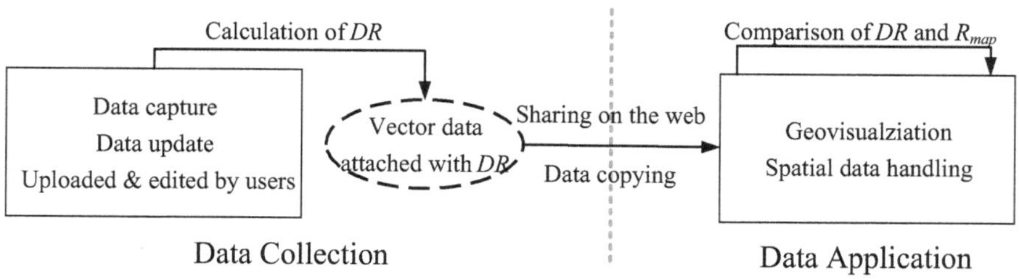

Fig. 3 The paradigm for computation and application of DR

Detecting LoD of OSM Based on Degree of Coalescence

OpenStreetMap(OSM) is the most famous VGI application. It is impossible to describe the *LoD* information of an OSM dataset by map scale because of the variety of data capture and imbalance of user expertise. This experiment tries to detect the *LoD* of OSM water dataset based on DR. we get OSM water dataset of Sichuan, china from OSM extraction API. This dataset's spatial reference is Web Mercator Projection based on WGS84 with about 3° spatial span over latitude and longitude. There are 198 features, 19,043 vertices in this dataset whose size is 738 KB in GeoJSON format.

Calculate the DR with the threshold of *DoC* 0.1and *cellsize*'s step size 1/10. As shown from the histogram of DR in Fig. 4, there are more than 80% features with DR ranging from 50 to 600 and is not a dominant DR with absolute superiority, therefore, it is unreasonable to use map scale to represent the *LoD* information of entire dataset anymore. Several features' DR are labeled in Fig. 5 with map resolution is 330. The number of features whose DR is bigger than 330 is 54, accounting for 27.2% of all features.

Detecting *LoD* of vector data based on DR has two advantages: (1) DR's calculation is device-independent and can be pre-computed using cloud computing offline. DR is a simple numeric value and can be stored as an attribute of geographic features. Meanwhile, high performance computing is preferable to improve the efficiency of DR algorithm. (2) DR can be calculated correctly even though the spatial reference, units and precision information are all missing. DR is robust and reliable enough for low quality geographic data sharing on the web.

Displaying OSM Data on Mobile Phone and Personal Computer (PC)

When drawing features on client-side (phone, computer, tablet, etc.), a feature's DR is compared with R_{map} of visualization to judge whether this feature needs

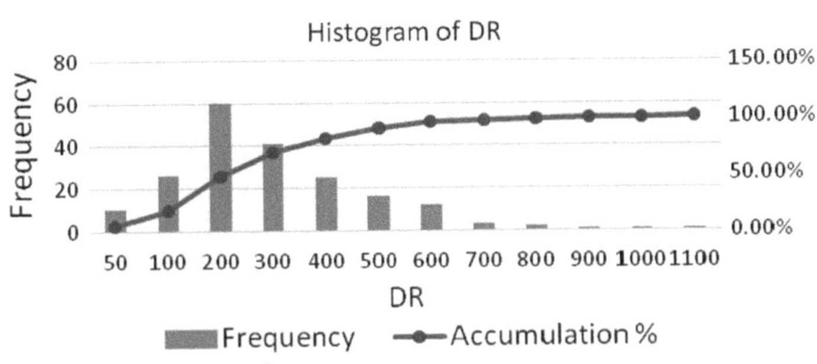

Fig. 4 Histogram of DR in OpenStreetMap water dataset

generalization or not. This process is efficient enough to operate real time. This section includes two experiments, one compares the difference between mobile phone and PC when displaying vector geographic data and another analyzes the influence of different symbol size on visualization.

Two devices' parameters are listed in Table 1. Visual distance of PC is about 500 mm, the minimum discernible size at this visual distance is 0.33 mm (SSC 2005), almost 1.2 pixels on PC monitor. Visual distance of mobile phone is 250 mm, the minimum discernible size at this visual distance is about 0.16 mm, nearly 2.5 pixels on mobile phone.

Experiment 1 compares the visualization of vector data between phone and PC. The data is OSM dataset used in experiment "Detecting LoD of OSM Based on Degree of Coalescence". Visualization on these devices are implemented based on HTML and JavaScript (Openlayers) which are supported by browsers on both devices. The initial R_{map} is 305.7. Without considering the output device, the number of features whose DR are bigger than 305.7 is 61, accounting for 30.8%. However, after taking device's size, resolution and visual distance into

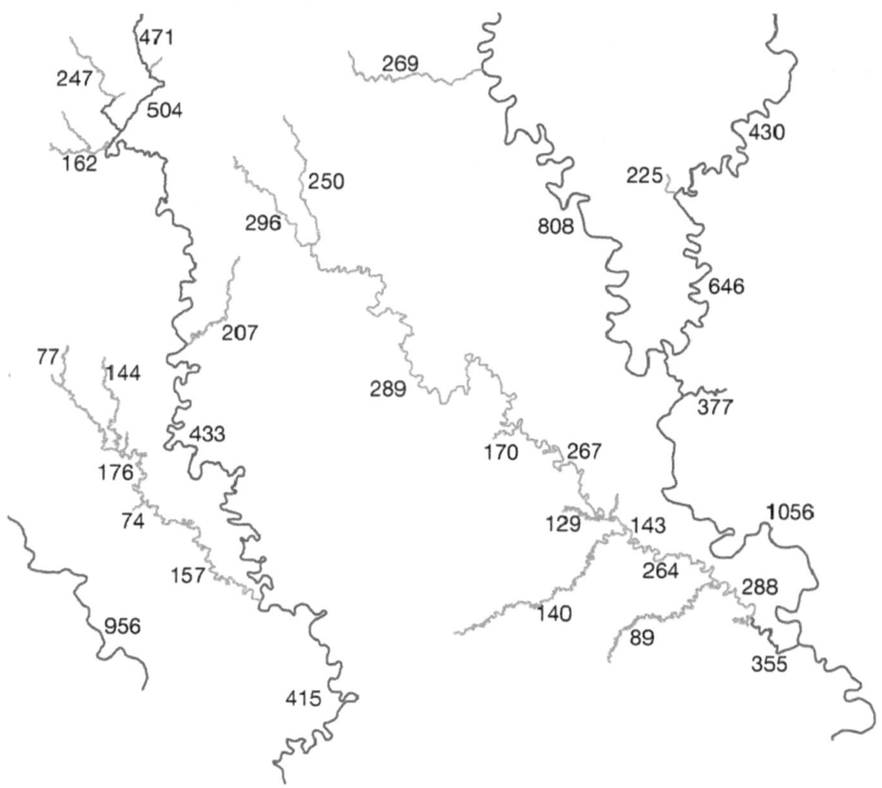

Fig. 5 Visual representation of DR in OpenStreetMap water dataset (Map resolution is 330. *Blue* means feature's DR < 330 and *red* means feature's DR >= 330)

Table 1 Parameters for devices with different size and same resolution

Device	Device resolution	Size (inch)	Pixel size (mm)	Visual distance (mm)	Discernible pixel number
Mobile Phone	1920 × 1080	5.7	0.065	250	2.5
PC	1920 × 1080	23.6	0.27	500	1.2

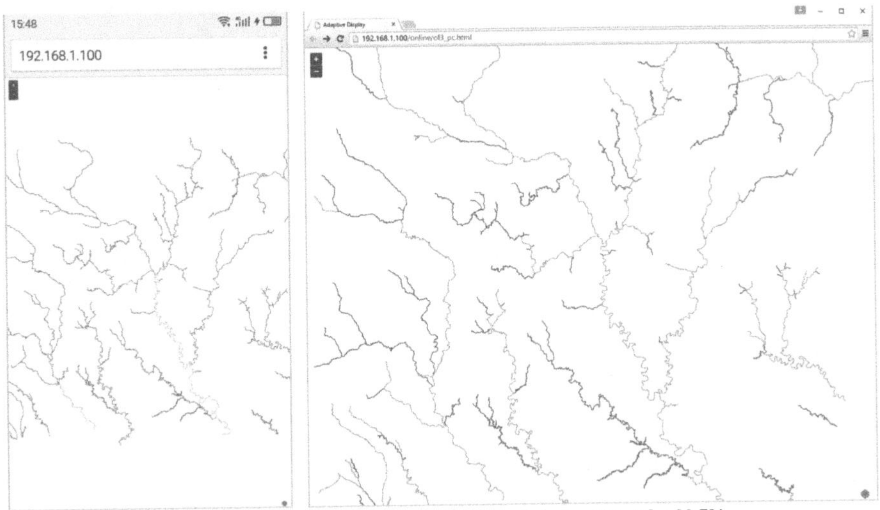

MP: Clear features account for **1.5%.** PC: Clear features account for **20.7%.**

Fig. 6 Visualization of OSM data on mobile phone and personal computer (*Red* denotes clear features and *blue* denotes unclear features)

consideration, there are fewer features meeting the constraint both on phone and PC. On mobile phone, legible feature's DR should be bigger than 305.7 × 2.5 = 764.25 and only 3 features are qualified (1.5%). On PC, legible feature's DR should be bigger than 305.7 × 1.2 = 366.84 and there are 41 features qualified (20.7%). Snapshots of two visualization results are listed in Fig. 6. In summary, constraint on visualization is more rigorous after taking device's parameters into consideration. Moreover, it is effective to distinguish features that are visual legible based on DR.

Experiment 2 compares symbolization's constraints on map visualization. As shown in Fig. 7, a few of features with 1 pixel line width in plot I become coalescent when line width is 2 pixels in plot II. These features (pointed at by gray arrows) are identified by filtering DR smaller than 305.7 × 1.2 × 2 but bigger than 305.7 × 1.2. Apparently, the results identified automatically are consistent with human perception.

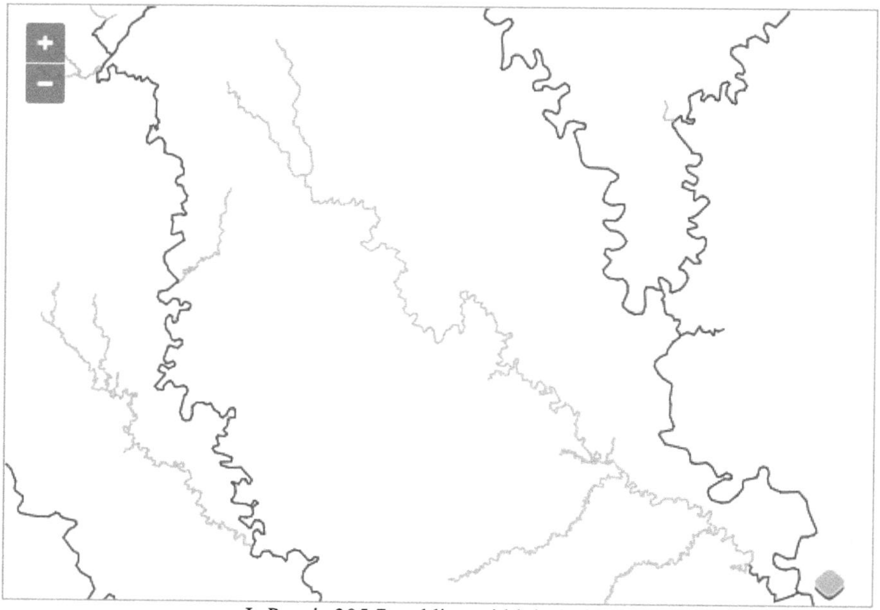

I. R_{map} is 305.7 and line width is 1 pixel.

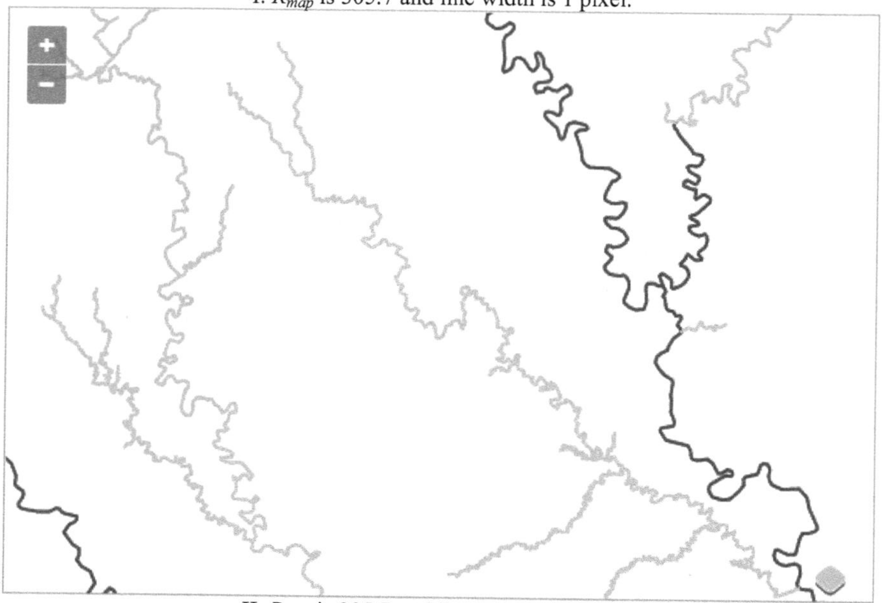

II. R_{map} is 305.7 and line width is 2 pixel.

Fig. 7 Different symbolization of OSM data on personal computer (*Red* denotes clear features and *blue* denotes unclear features)

Conclusion

In this paper, we proposed a new model called **Detail Resolution** to Describe Level of Detail Information of Vector GIS Data. The proposed DR model is specific for digital display and has several advantages listed below:

(1) The *DoC* measure can be used to detect the *LoD* information of vector data without metadata, no matter what the data's unit, spatial reference, precision, etc. is.
(2) Detail Resolution can provide fine-grained *LoD* information at the feature level. Each feature has a unique number to indicate its *LoD*, which can deal with the heterogeneity of VGI effectively.
(3) Detail Resolution is expressed by a simple numeric value which is easily stored as one attribute of a geographic feature. This numeric value should be calculated beforehand and updated with the change of geometry information.
(4) This model supports adaptive scale transformation of vector data for geospatial visualization. The DR number is transferred to the client together with the data and it is easy to identify which feature is coalescent through simple comparison between DR and actual size of MBR in pixels. This paradigm supports the new form of map visualization integrating offline preprocessing and online scale transformation effectively and efficiently.

The future work will try to extend this model to area features and point features. Furthermore, the DR index proposed only represent one dimension of *LoD* information and incorporating other indicators to build a complete *LoD* model is another future challenge.

References

Hecht, R., Kunze, C., & Hahmann, S. (2013). Measuring completeness of building footprints in OpenStreetMap over space and time. *ISPRS International Journal of Geo-Information, 2*(4), 1066–1091.

Li, Z., & Openshaw, S. (1993). A natural principle for the objective generalization of digital maps. *Cartography and Geographic Information Systems, 20*(1), 19–29.

Mei, Y., Li, L., & He, B. (2008). Cartographic visualization based on boundary anti-aliasing. *Geomatics and Inforamtion Science of Wuhan University, 33*(7), 759.

OGC. (2011). OpenGIS® Implementation Standard for Geographic information—Simple feature access—Part 1: Common architecture OGC 06–103r4.

Shea, K. S., & McMaster, R. B. (1989). Cartographic generalization in a digital environment: When and how to generalize. In *Proceedings Auto-Carto 9* (pp. 56–67).

SSC. (2005). *Topographic maps—map graphics and generalization: Swiss Society of Cartography.*

Touya, G., & Brando-Escobar, C. (2013). Detecting level-of-detail inconsistencies in volunteered geographic information data sets. *Cartographica: The International Journal for Geographic Information and Geovisualization, 48*(2), 134–143.

Part IV
Spatial Analysis and Simulation

Urban Growth Evaluation: A New Approach Using Neighborhood Characteristics of Remotely Sensed Land Use Data

Shyamantha Subasinghe and Yuji Murayama

Introduction

The physical and functional transformation of rural landscapes into urban forms is recognized as urban growth (Thapa and Murayama 2010). According to Clark (1982), urban growth is a spatial and demographic process that is characterized by a change in population distribution from a village to a town or city. Currently, rapid urban growth is a major worldwide trend, involving a variety of resources and environmental problems, such as habitat loss, species extinction, land-cover change, and alteration of hydrological systems (Hahs et al. 2009; Jain 2011). Driven by this trend, the understanding of urbanization has pushed to the forefront of environmental and development agendas (Mertes et al. 2015).

The typical spatial organization of individual urban areas is explained in von Thünen's (1826) bid-rent theory, Burgess's (1925) concentric zone model, Christaller's (1933) central place theory, and Hoyt's (1939) sector model. Although these studies have formed foundations for subsequent work, they are predominantly descriptive models that assume cities grow in a uniform or linear manner, and most do not contribute to the understanding of the spatiotemporal patterns of urban forms or growth (Dietzel et al. 2005). In addressing this limitation, various new and sophisticated methods have been developed and successfully applied for charac-

S. Subasinghe (✉) · Y. Murayama
Graduate School of Life and Environmental Sciences, University of Tsukuba,
1-1-1 Tennodai, Tsukuba City, Ibaraki 3058572, Japan
e-mail: subasinghepgia@gmail.com

Y. Murayama
e-mail: mura@geoenv.tsukuba.ac.jp

C. Zhou et al. (eds.), *Spatial Data Handling in Big Data Era*,
Advances in Geographic Information Science,
DOI 10.1007/978-981-10-4424-3_13

terizing urban growth. Batty and Longley (1994) considered urban growth a cellular fractal stochastic process and modeled urban growth through the cellular automata method. Dietzel et al. (2005) suggested that the process of urban growth could be characterized through diffusion and coalescence. To connect the theory of diffusion and coalescence, three indicators of urban growth patterns—infill, extension, and leapfrog development—have also been identified (Estoque and Murayama 2015). In addition, in urban planning initiatives, the importance of the low, moderate, and high sublevels of each indicator has been highlighted.

By addressing the lack of more detailed urban growth identification, the remote sensing of urban landscapes has recently led to a number of new approaches to characterize urban growth on various spatial scales (Antrop 2004; Kantakumar et al. 2016; Xian and Crane 2005). Among them, ULU change analysis with a spatial metric has been widely applied (Aguilera et al. 2011; Estoque and Murayama 2011). Two major methods of land change analysis developed are spectrally based (image-to-image) and classification-based (map-to-map) change detections (Xian and Crane 2005). Furthermore, a large volume of successful research studies has employed both of these methods when characterizing ULU change in general and urban growth in particular (Dorning et al. 2015; Guindon et al. 2004; Mertes et al. 2015).

However, remote sensing applications for the urban growth evaluation still pose several limitations. Fundamentally, inconsistency in ULU definitions has created challenges in urban growth detection and evaluation (Taubenböck et al. 2012). Due to this inconsistency, remote sensing studies typically describe built environments as ULU, and the non-built environments as non-urban land use (Estoque and Murayama 2015; Liu et al. 2016; Su et al. 2011). However, some non-built land use dominates urban areas (e.g., parks and runways) in reality, and function as ULU. Thus, characterizing the urban area using only the built environments confound our understanding of urban growth (Mertes et al. 2015). In such a context, characterizing ULU classification based on their locational contexts or neighborhood interaction is vital and helps us to detect urban growth in a more realistic manner. Moreover, the neighborhood interaction of a surrounding area can be employed to elucidate low, moderate, and high levels of urban growth by determining major patterns.

In general, morphological spatial pattern analysis (MSPA) allows the integration of neighborhood interaction in defining ULU categories and helps to determine the levels of urban growth in a contextual manner (Ostapowicz et al. 2008; Vogt et al. 2007). Using MSPA, Vogt et al. (2007) developed a forestland classification (e.g., core, patch, perforated, and edge) based on forest and non-forest land categories. Angel et al. (2010) developed an urban land classification (urban, suburban, rural, fringe open space, exterior open space, and rural open space) based on built and non-built land categories. They have employed only binary land classification to classify the forest- related or urban- related land use categories. However, developing ULU classifications using binary land use or cover categories may be insufficient due to existence of a higher complexity of ULU (Jiao 2015; Zhou et al. 2015). In such a context, incorporation of ancillary data and multiple land categories with MSPA will provide more advancement in ULU classification and a clearer understanding of growth patterns.

In this study we present a new approach to recognize the spatial pattern of urban growth by integrating the neighborhood interactions of ULU categories. We called our approach the Urban Growth Evaluation Approach (UGEA); it was tested using a case study of the Colombo metropolitan area, Sri Lanka.

Concept of Neighborhood Interaction

Neighborhood interactions are an important component of many land use models connecting to the Tobler's (1970) first law of Geography ("Everything is related to everything else but near things are more related than distance things"). Cellular automata (CA) is commonly used to implement neighborhood interactions in land use models through Vin Neumann's adjacent four cells rule (Fig. 1a) or Moor's adjacent eight cells rule (Fig. 1b). In reality, a cell does not only influence the state of adjacent cells but also those located at a certain distance, although with less effect (Barreira-González et al. 2015). In this respect, distance decay function can be used to integrate neighborhood interaction to the cells (Fig. 1c) (Zhao and Murayama 2011).

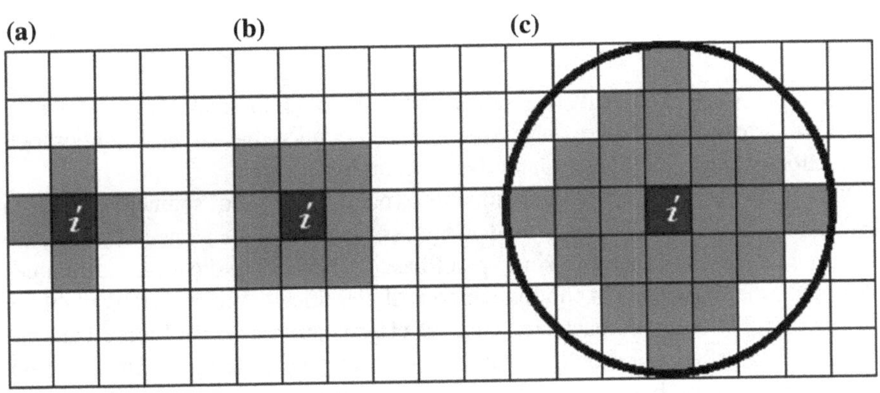

Fig. 1 **a** Van Neumann's concept (4 cells), **b** Moore's concept (8 cells), and **c** distance decay concept (i = processing cell)

Methods

Data

To test our UGEA, we acquired Landsat TM/ETM + images from the United State Geological Survey (USGS) website during the study area's wet seasons in 2001 and 2014. One Landsat scene (path 141 and row 55) covering the entire study area was collected for these two time points. The two images collected were Landsat-7 ETM + of 2001 and Landsat-8 OLI/TIRS of 2014. The Landsat-7 ETM + of 2001 and Landsat-8 OLI/TIRS images were Standard Terrain Correction (L1T) (Taubenböck et al. 2012) and cloud free. Therefore, geometric correction and atmospheric corrections were not preformed. In addition to Landsat images, Google Earth™ images and topographical maps (Department of Survey, Sri Lanka) were used for accuracy assessment and to delineate boundaries of some land use (i.e., protected areas, runways, etc.).

Determining Spatial Patterns of Urban Growth

Basically, UGEA turns the ULU change maps into an urban growth map through the several processes. All the processes can be summarized into three major steps (Fig. 2): (1) ULU mapping, (2) identification major spatial patterns of urban growth, and (3) development of sublevels of urban growth.

ULU Mapping

As the first step of UGEA, we developed a method to map ULU rationally. The ULU categories in the maps were mainly defined based on neighborhood interactions of the study area's land use categories.

We employed the hybrid classification (pixel-based and segment-based) to develop the initial study area's land use classification. With a method, first, we classified Landsat images using pixel-based (PB) classification techniques employing the maximum likelihood supervised classification approach available in the ENVI 5.2™ software package. This PB classification produced three land use categories: built (meaning built-up lands), non-built (meaning non-built up lands), and water (meaning bodies of water). Second, we classified the study area's land uses using segment-based (SB) classification. In SB classification, Landsat images were segmented using the ENVI 5.2 software package and produced two land uses: protected areas, and urban open space (runways, playgrounds, and parks), employing region merging techniques of SB classification technique. The SB classification method is the most appropriate classification method to classify land

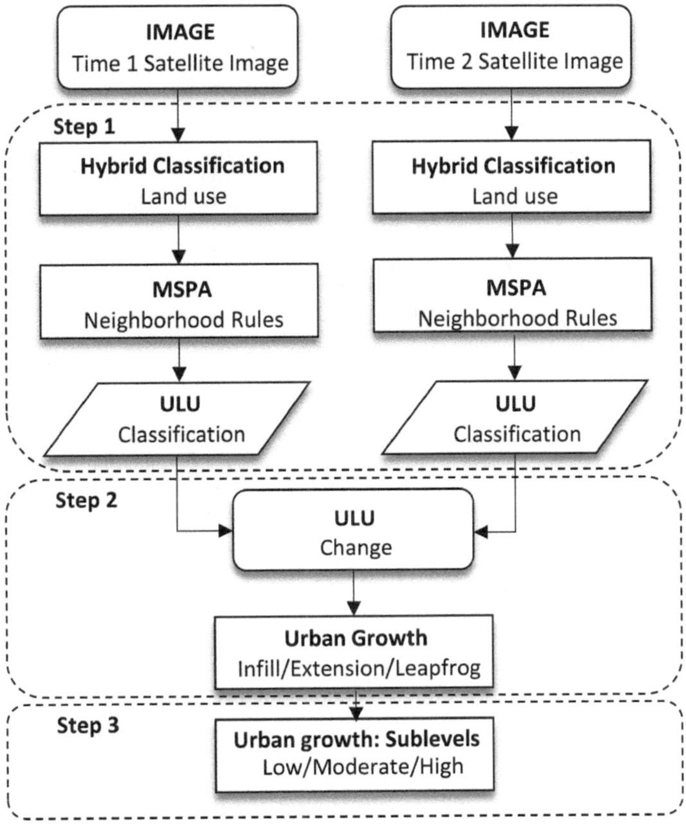

Fig. 2 Flowchart showing all the steps in UGEA

use categories having specific boundaries or edges such as parks, protected areas etc. (Blaschke 2010). Third, the results of PB classification and the SB classification were integrated using the raster algebra tool in the ArcGIS™ software package. The final output of this hybrid classification contained five land use categories: built, non-built, protected areas, urban open spaces (parks, playgrounds, and runways), and water. It is agreed that a higher level of accuracy can be maintained with the hybrid land use classification method than individual PB or SB classification methods (Li et al. 2013). The accuracy of land use classification was checked using 300 samples at each time points (2001 and 2014) through careful and rigorous visual inspection. Google Earth images were used as reference data for accuracy assessment and the overall accuracy was (Congalton 1991) 90.33, and 92.66% for 2001, and 2014 respectively in this hybrid classification method.

Neighborhood interaction rules were processed using MSPA to convert the study area's land use into ULU mapping. To process the neighborhood rules, we first defined the active land use categories and inactive land use categories in the study area. The active land use category means the land use categories that influence ULU

Table 1 Neighborhood interaction rules of ULU categories

ULU categories	Description of neighborhood interaction rule
Urban dense	50–100% built-up pixels in a 1-km^2 area of neighborhood: Buffer with 564 m map unit (18 pixels) distance from built pixel was employed to determine a 1-km^2 area
Urban sparse	10–50% built-up pixels in a 1-km^2 area of neighborhood: Buffer with 564 meters map unit (18 pixels) distance from built pixel was employed to determine a 1-km^2 area
Urban open space	Non-built land within a 100-m distance from urban area: Buffer with 100 meters map units (3 pixels) distance from urban built was employed to determine a 1-km^2 area
Captured urban open space	Patches of non-built, less than 2 km^2, completely surrounded by urbanized area (included urban dense, urban sparse, and urban open space)
Urban fringe	100-m (3 pixels) distance edge in between urbanized (included urban dense, urban sparse, and urban open space) and non-urban area (included non-urban built and non-urban open space)
Non-urban built	0–10% built up pixels in a 1-km^2 area: buffer with 564 m map unit (18 pixels) distance from built pixel was employed to determine a 1-km^2 area
Non-urban open space	All other land use

Note All the measures are computed based on raster data 30 m × 30 m pixels

classification as a neighborhood. Inactive land use means the land use categories that do not influence ULU classification as a neighborhood. Here, we considered built and non-built land categories as active land use categories and protected areas, urban open space (parks, playgrounds), and water as inactive land use categories. Second, the neighborhood interaction rules (Table 1) were processed and defined ULU categories. The neighborhood rules were performed for each pixel of land use using the urban growth analysis (UGA) tool, developed by the Center for Landuse Education and Research Institute (CLER), and the ArcGIS focal analysis tool. As a result of this process, seven ULU categories (urban dense, urban sparse, urban open space, captured urban open space, urban fringe and non-urban area) were classified.

Figure 3 illustrates how the neighborhood interaction rules are processed on a cell space.

Later, these seven ULU were integrated with protected areas, urban open spaces, and water, which are classified in hybrid classification. The protected area was converted into non-urban open space and the final ULU map was contained eight categories: urban dense, urban sparse, urban open space, captured urban open space, urban fringe, non-urban built, non-urban open space, and water. In the present study, we produced two ULU maps with eight categories for 2001 and 2014 to detect the spatial patterns of urban growth.

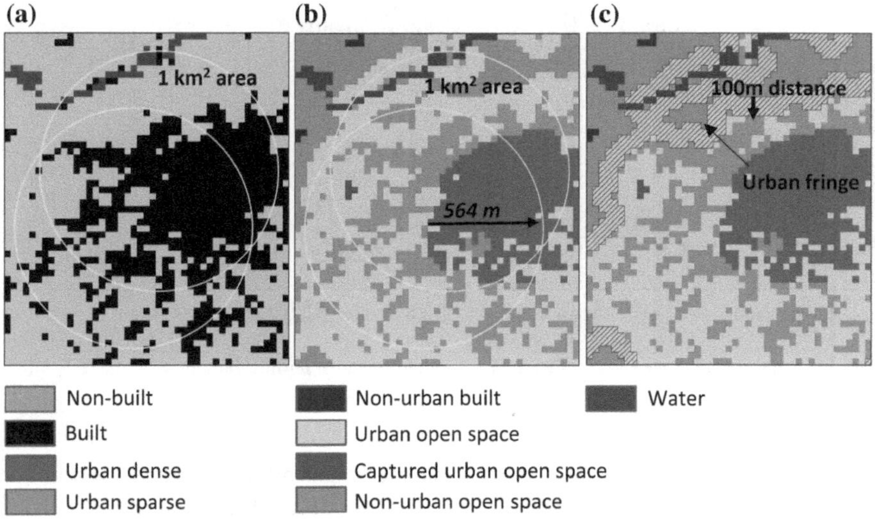

Fig. 3 The neighborhood interaction area: **a** and **b** the percentage of built pixels is calculated within a 1 km^2 area, and **c** distance from non-urban areas is determined

Identification of Major Spatial Pattern of Urban Growth

To spatially characterize urban growth, we distinguished the three major spatial patterns of urban growth—infill, extension, and leapfrog development. The ULU transition from the initial time point and the final time point were used to detect these growth patterns (Table 2).

Briefly, each urban growth pattern contains the following characteristics: (1) infill, characterized by new urban development that occurs in an already

Table 2 ULU changes used to characterize the three major urban growth patterns

Urban growth pattern	Change from	Change to
Infill	Urban open space	Urban dense
	Urban open space	Urban sparse
	Captured urban open space	Urban dense
	Captured urban open space	Urban sparse
Leapfrog	Non-urban built	Urban dense
	Non-urban built	Urban sparse
	Non-urban open space	Urban dense
	Non-urban open space	Urban sparse
Extension	Any above transition occurs in the urban fringe area and connected new development to the extension	

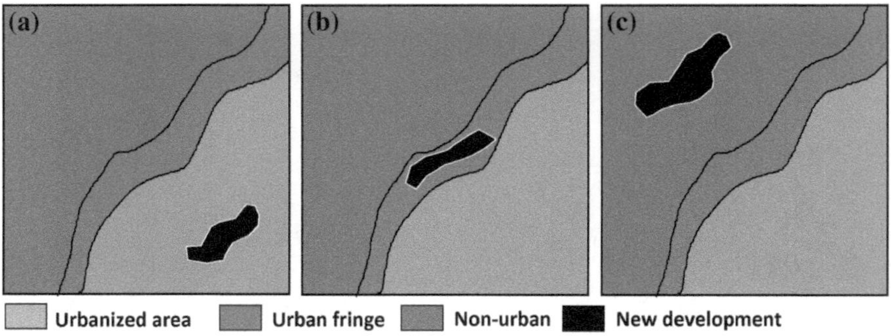

Fig. 4 Locational characteristics of each development pattern: **a** infill, **b** extension, and **c** leapfrog development

urbanized area (Fig. 4a); (2) extension, characterized by new urban development which occurs in the urban fringe area and connects it to new development, (Fig. 4b); (3) leapfrog development, characterized by new development that occurs in a non-urban area (Fig. 4c).

Concept of Urban Growth Sublevels

We further separated the major patterns of urban growth into three sublevels: low level, moderate level, and high level. These sublevels were determined based on the nature of the development in the surrounding area or neighborhood interaction. In doing so, the nature of urban dense land category and urban sparse land category was considered within a 1-km^2 area (same as ULU classification). A buffer with 564 m of distance was employed to delineate the area of neighborhood interaction (a 1-km2 area). Figure 4 illustrates examples for locational characteristics of each sublevel separation.

Figure 5a illustrates the urban growth occurring in an area where the surrounding area is characterized by a low level of development. Figure 5b illustrates urban growth occurring in an area where the surrounding area is characterized by a moderate level of development. Figure 5c illustrates urban growth occurring in an area where the surrounding area is characterized by a high level of development.

Sublevel Separation Process

We employed the Map algebra tool in ArcGIS to calculate the proportion of urban dense area and urban sparse area as a percent of the total land area (except water) within a 1-km^2 area. In this processing, two main raster layers were used. Figure 6

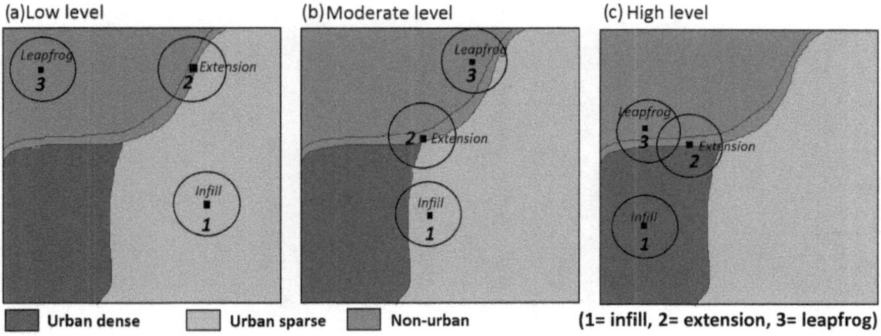

Fig. 5 The sublevels of urban growth patterns

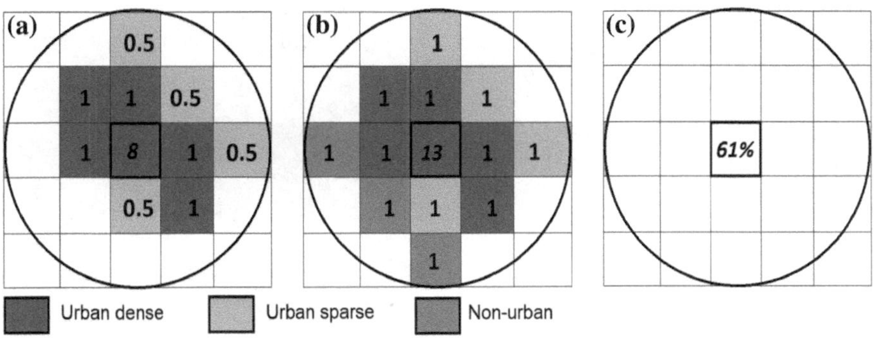

Fig. 6 Calibrated values of land use: **a** first layer containing only urban dense and urban sparse areas, **b** second layer containing all land uses (except water), and **c** resulting layer with percentage values

illustrates a simplified example the calibration process used for sublevel separation. The first layer containing only urban dense and urban sparse calibrated was calibrated (urban dense value = 1, and urban sparse = 0.5). The second layers contain ULU categories, which were calibrated with ULU value = 1, except water (water = no data). The percentage of the first layer was calculated according to the presence of the second layer. This calculation is simply explained in Eq. 1.

$$Sl = \frac{\sum(Dp + Sp)}{UL} \times 100 \quad (1)$$

where *Sl* is the percentage of the development level of the surrounding area, *Dp* is the total value of urban dense pixels in the first layer, *Sp* is the total value of urban sparse pixels in the first layer, and *UL* is the total value of ULU categories in the second layer.

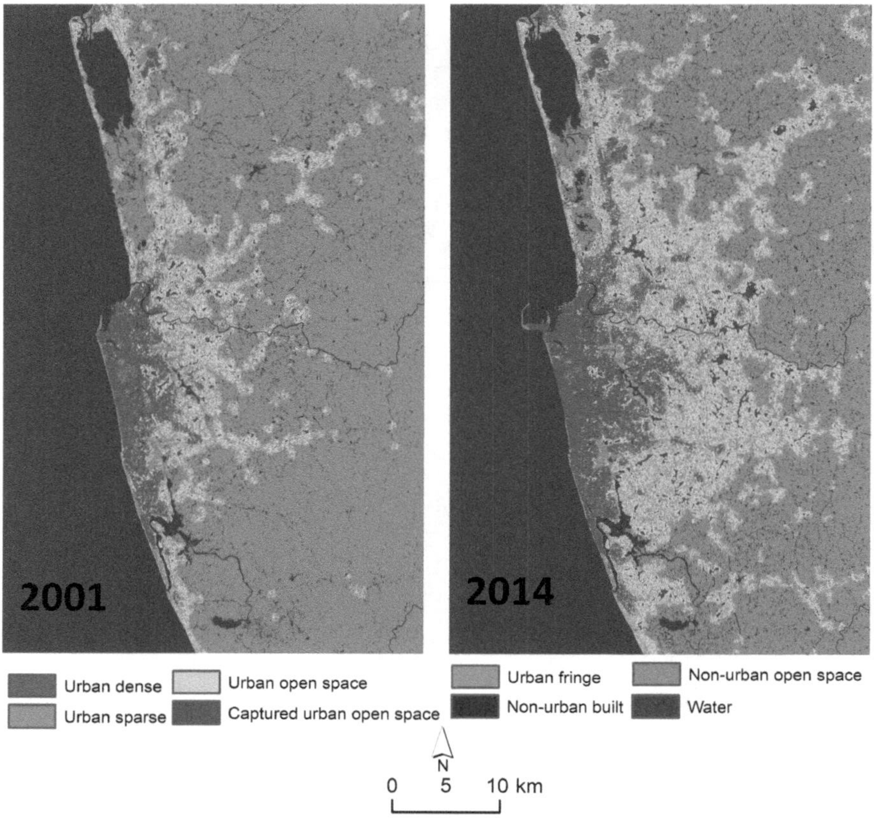

Fig. 7 ULU maps for 2001 and 2014

Depending on the produced percentage of development for each pixel, the main patterns of urban growth (infill, extension, and leapfrog development) were subdivided into low level (0–20%), moderate level (20–70%), and high level (70–100%).

Results

Figure 7 presents the results of ULU mapping for 2001 and 2014.

The major spatial patterns of urban growth, derived from the ULU change, are presented in Fig. 8a, and the sublevels of each pattern are presented in Fig. 8b.

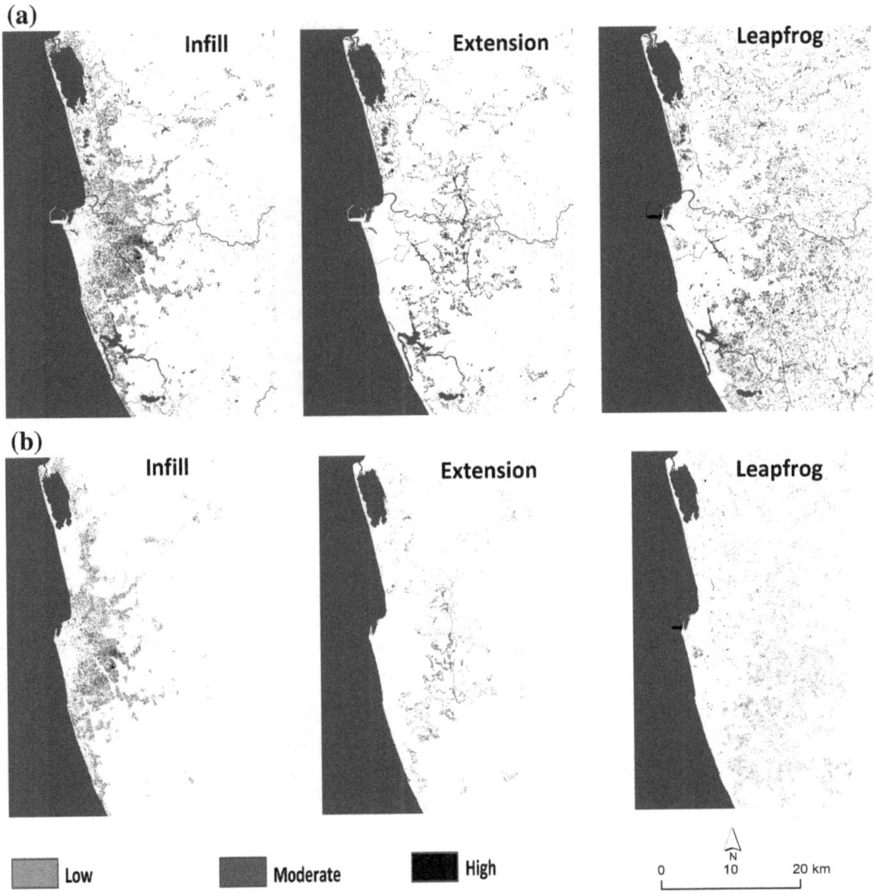

Fig. 8 The urban growth patterns: **a** major patterns, and **b** sublevels

Discussion

UGEA as a Geographic Approach

Fundamentally, an approach contains a set of methods and techniques with clear starting points, transition points, and end points. Geographic approaches allow us to understand the world by organizing, analyzing, and modeling various geographic data (Dangermond 2007).

McHarg (1969) articulated the philosophical context of the geographic approach for managing human activities within natural and cultural landscapes. His approach created a fundamental factor for geographers to analyze our world. The geographic approach consists of five steps (Fig. 9).

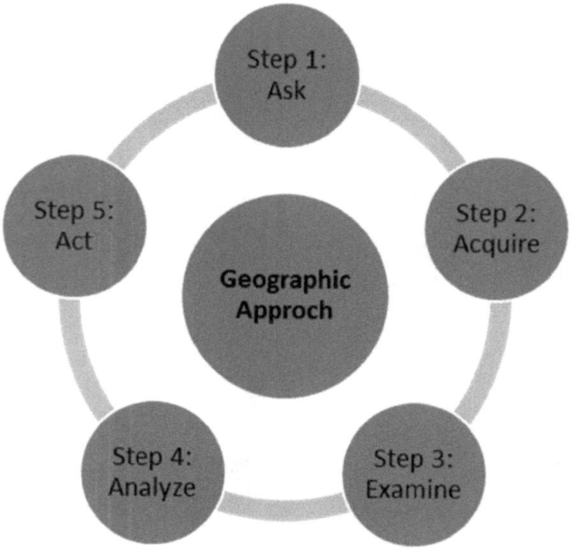

Fig. 9 Five steps of the geographic approach

The first step formulates the question from a location-based perspective. In present study, as the research question, we ask, "What are the spatial patterns of urban growth?" This question establishes the urban environment as the geographic context as we attempt to understand the spatial patterns of urban growth. Analyses integrating spatial aspects lead to a greater fundamental understanding of the dynamic process involved and thereby aid in the development of actual solutions (Ding and Elmore 2015).

The second step determines the necessary data that should be acquired by the analysis. We deal with remotely sensed big data in a data-poor environment. Satellite images are the main data and are to produce ULU maps by image processing. Satellite image processing comprises four basic operations: (1) image restoration, (2) image enhancement, (3) image classification, and (4) image transformation (Thompson et al. 2002). In the present study, we mainly conducted geometric corrections and visual enhancement for Landsat images, and image processing with MSPA in relation to this step.

The third step examines the acquired data to understand whether the prepared data is appropriate for achieving the objectives and answering the research questions. It is necessary to visually inspect it and understand how the data is organized. Here, we visually inspected and assessed the accuracy of our outputs.

The fourth step performs the data analysis. After examining the data, here we analyzed the spatiotemporal pattern of urban growth and separated it into sublevel based on its neighborhood interaction. Furthermore, the difference in urban growth was analyzed based on time intervals.

The fifth step presents the results visually. Visual presentation through maps, tables, and charts is a common method with the geographic approach. The International Cartographic Association's research agenda identified four visualization goals: exploration, analysis, synthesis, and presentation (MacEachren 1994). In this study, we present our ULU classification results and urban growth pattern visually using geospatial techniques.

Contributions of the UGEA

The problem of identifying a more realistic means of urban space representation and urban growth identification appears to have been almost solved by high resolution remote sensing imagers in the big data era (Barreira-González et al. 2015). However, the practical applications of high resolution satellite images to characterize the urban growth of large urban areas such as metropolitan areas (particularly in developing countries) have been limited by the cost and availability of high resolution satellite imagery. Similarly, the lack of socioeconomic data in developing countries has also limited the urban growth evaluation.

There has been rapid and vital growth of urban areas in developing countries over the last two decades, and drastic urban growth is predicted for these regions in the future (Cohen 2006; Seto and Fragkias 2005). Thus, the main purpose of the present study was to develop a new approach to characterize the spatiotemporal patterns of urban growth with minimal data input and complexity for widespread use and applications. The introduced UGEA can be performed with Landsat imagery and widely available ancillary data (i.e., Google Earth images, and topographical maps). Because of this advantage, the application of this approach in developing countries can be assured.

As previously mentioned, earlier approaches, which employed limited sources, mostly used the built-up areas as urban areas, and urban growth was characterized using the land use change from non-built to built. Our proposed UGEA uses the advantages of the neighborhood interaction concept to overcome the narrow view of previous studies, and introduced a wide range of urban land use categories and urban growth patterns. In this sense, the UGEA enables a conceptual and practical solution to characterize urban areas in a data-poor context.

Although the neighborhood interaction concept with remotely sensed land use may be a good option, it also presents some limitations. An urban area not only depends on the neighborhood land use types, but also on socioeconomic and political factors, which are highly influential in the urban areas (Kantakumar et al. 2016). Thus, it is necessary to integrate these factors with remotely sensed data to define an urban area. Furthermore, this study remained "blind to pattern" (Longley 2002) and it requires a knowledge of processes and driving forces to characterize urban areas in a more comprehensive manner.

Technically, we analyzed urban growth patterns in the study area using ArcGIS focal analysis and the UGA tool. The processing of this big data with neighborhood interaction rules in the ArcGIS environment is very time consuming and costly; therefore the use of Python code, a one of the widespread programming languages in geospatial analysis and data management may be, an appropriate solution.

Conclusions

The new approach introduced in this study—UGEA—addresses two key urban application needs. As a key to urban growth evaluation, the UGEA initially develops land use classification using the neighborhood interaction rules of land use. In general, ULU classification is associated with several difficulties related to medium resolution satellite imagery like Landsat due to the higher level of complexity and heterogeneity of urban areas. Thus, the classification of ULU categories from Landsat requires knowledge of the larger scale of the spatial context. The concept of the neighborhood that is available in geospatial analysis enabled a solution to incorporate a large-scale spatial context to our ULU mapping.

Subsequently, urban growth was detected using three patterns (infill, extension, and leapfrog development) and separated into different levels depending on the locational context. The incorporation of location context to the sublevel classification of urban growth is a new idea introduced in this study; it can be further developed in the future. Here, we used only the distribution of urban dense and urban sparse land categories to separate the sublevels, but the incorporation of additional urban features (i.e., industries, administrative, and services) can lead to more sophisticated sublevel classification.

This approach is more applicable to comparative urban studies than to individual case studies. Comparative analysis would help to elucidate the urbanization process of each city separately and compare the difference in urbanization processes. In such a context, the development of a GIS-based tool to conveniently run all the steps of UGEA would be useful in future research activities.

Acknowledgements This study was supported by the Japan Society for the Promotion of Science (Doctoral Fellowship Grant: ID No. 15J00611, 2015–16; and Grant-in-Aid for Scientific Research B: No. 26284129, 2016-16, Representative: Shyamantha Subasinghe). The first author gratefully acknowledges Prof. Jason Parent, University of Connecticut for providing UGA tools and the Python code.

References

Aguilera, F., Valenzuela, L. M., & Botequilha-Leitão, A. (2011). Landscape metrics in the analysis of urban land use patterns: A case study in a Spanish metropolitan area. *Landscape and Urban Planning, 99*(3–4), 226–238.

Angel, S., Parent, J., & Civco, D. L. (2010). *The fragmentation of urban footprints: Global evidence of sprawl, 1990–2000*. Lincoln Institute of Land Policy Working Paper: 1–100.

Antrop, M. (2004). Landscape change and the urbanization process in Europe. *Landscape and Urban Planning, 67*(1–4), 9–26.

Barreira-González, P., Gómez-Delgado, M., & Aguilera-Benavente, F. (2015). From raster to vector cellular automata models: A new approach to simulate urban growth with the help of graph theory. *Computers, Environment and Urban Systems, 54,* 119–131.

Batty, M., & Longley, P. A. (1994). *Fractal cities: A geometry of form and function* (1st ed.). San Diego, CA: Academic press.

Blaschke, T. (2010). Object based image analysis for remote sensing. *ISPRS Journal of Photogrammetry and Remote Sensing, 65*(1), 2–16.

Burgess, E.W. (1925). The growth of the city: An introduction to a research project. In R. E. Park, E. W. Burgess & R. D. McKenzie (Eds.), *The city* (pp. 47–62). Chicago: Chicago University Press.

Christaller, W. (1933). *Central places in Southern Germany* (C.W. Bakin, Trans, 1966). New Jersey, USA: Prince Hall.

Clark, D. (1982). *Urban geography: An introductory guide* (1st ed.). London: Croom Helm.

Cohen, B. (2006). Urbanization in developing countries: Current trends, future projections, and key challenges for sustainability. *Technology in Society, 28*(1–2), 63–80.

Congalton, R. G. (1991). A review of assessing the accuracy of classifications of remotely sensed data. *Remote Sensing of Environment, 37*(1), 35–46.

Dangermond, J. (2007). *GIS: The geographic approach.* Retrieved May 25, 2016, from http://www.esri.com/news/arcnews/fall07articles/gis-the-geographic-approach.html

Dietzel, C., Oguz, H., Hemphill, J. J., Clarke, K. C., & Gazulis, N. (2005). Diffusion and coalescence of the Houston Metropolitan Area: Evidence supporting a new urban theory. *Environment and Planning B: Planning and Design, 32*(2), 231–246.

Ding, H., & Elmore, A. J. (2015). Spatio-temporal patterns in water surface temperature from Landsat time series data in the Chesapeake Bay, U.S.A. *Remote Sensing of Environment, 168,* 335–348.

Dorning, M. A., Koch, J., Shoemaker, D. A., & Meentemeyer, R. K. (2015). Simulating urbanization scenarios reveals tradeoffs between conservation planning strategies. *Landscape and Urban Planning, 136,* 28–39.

Estoque, R. C., & Murayama, Y. (2011). Spatio-temporal urban land use/cover change analysis in a hill station: The case of Baguio city, Philippines. *Procedia Social and Behavioral Sciences, 21,* 326–335.

Estoque, R. C., & Murayama, Y. (2015). Intensity and spatial pattern of urban land changes in the megacities of Southeast Asia. *Land Use Policy, 48,* 213–222.

Guindon, B., Zhang, Y., & Dillabaugh, C. (2004). Landsat urban mapping based on a combined spectral-spatial methodology. *Remote Sensing of Environment, 92*(2), 218–232.

Hahs, A. K., McDonnell, M. J., McCarthy, M. A., Vesk, P. A., Corlett, R. T., Norton, B. A., & Williams, N. S. G. (2009). A global synthesis of plant extinction rates in urban areas. *Ecology Letters, 12*(11), 1165–1173.

Hoyt, H. (1939). *The structure and growth of residential neighborhoods in American cities.* Washington: Federal Housing Administration.

Jain, M. (2011). A next-generation approach to the characterization of a non-model plant transcriptome. *Current Science, 101*(11), 1435–1439.

Jiao, L. (2015). Urban land density function: A new method to characterize urban expansion. *Landscape and Urban Planning, 139,* 26–39.

Kantakumar, L. N., Kumar, S., & Schneider, K. (2016). Spatiotemporal urban expansion in Pune metropolis, India using remote sensing. *Habitat International, 51,* 11–22.

Li, X., Meng, Q., Xingfa, G., Jancso, T., Yu, T., Wang, K., et al. (2013). A hybrid method combining pixel-based and object-oriented methods and its application in Hungary using Chinese HJ-1 satellite images. *International Journal of Remote Sensing, 34*(13), 4655–4668.

Liu, Y., He, Q., Tan, R., Liu, Y., & Yin, C. (2016). Modeling different urban growth patterns based on the evolution of urban form: A case study from Huangpi, Central China. *Applied Geography, 66,* 109–118.

Longley, P. A. (2002). Geographical information systems: Will development in urban remote sensing and GIS lead to 'better' urban geography? *Progress in Human Geography, 26,* 231–239.

McEachren, A. M. (1994). Visualization in modern cartography: Setting the agenda. In A. M. Maceachren & D. R. F. Taylor (Eds.), *Visualization in modern cartography* (pp. 1–13). Oxford: Pergamon.

Mcharg, I. L. (1969). Design with nature. *Design with Nature,* 1–16.

Mertes, C., Schneider, A., Sulla-Menashe, D., Tatem, A., & Tan, B. (2015). Detecting change in urban areas at continental scales with MODIS data. *Remote Sensing of Environment, 158*(158), 331–347.
Ostapowicz, K., Vogt, P., Riitters, K. H., Kozak, J., & Estreguil, C. (2008). Impact of scale on morphological spatial pattern of forest. *Landscape Ecology, 23*(9), 1107–1117.
Seto, K. C., & Fragkias, M. (2005). Quantifying spatiotemporal patterns of urban land-use change in four cities of China with time series landscape metrics. *Landscape Ecology, 20*(7), 871–888.
Su, S., Jiang, Z., Zhang, Q., & Zhang, Y. (2011). Transformation of agricultural landscapes under rapid urbanization: A threat to sustainability in Hang-Jia-Hu region, China. *Applied Geography, 31*(2), 439–449.
Taubenböck, H., Esch, T., Felbier, A., Wiesner, M., Roth, A., & Dech, S. (2012). Monitoring urbanization in mega cities from space. *Remote Sensing of Environment, 117,* 162–176.
Thapa, R. B., & Murayama, Y. (2010). Drivers of urban growth in the Kathmandu valley, Nepal: Examining the efficacy of the analytic hierarchy process. *Applied Geography, 30*(1), 70–83.
Thompson, M., Gonzalez, R. C. R., Wintz, P., Woods, R. E. R., & Masters, B. R. (2002). Digital image processing. *Leonardo* (Vol. 14).
Vogt, P., Riitters, K. H., Estreguil, C., Kozak, J., Wade, T. G., & Wickham, J. D. (2007). Mapping spatial patterns with morphological image processing. *Landscape Ecology, 22*(2), 171–177.
Von Thünen, J. (1826). *The isolated state (English version).* London: Pergamon.
Xian, G., & Crane, M. (2005). Assessments of urban growth in the Tampa Bay watershed using remote sensing data. *Remote Sensing of Environment, 97*(2), 203–215.
Zhao, Y., & Murayama, Y. (2011). Modelling neighborhood interaction in cellular automata-based urban geosimulation. In Y. Murayama & R. P. Thapa (Eds.), *Spatial analysis and modeling geographical transformation process* (pp. 75–87). New York: Springer.
Zhou, N., Hubacek, K., & Roberts, M. (2015). Analysis of spatial patterns of urban growth across South Asia using DMSP-OLS nighttime lights data. *Applied Geography, 63,* 292–303.

Influential Factors of Building Footprint Location and Prediction of Office Shape in City Blocks in Tokyo's Commercial Zones

Masahiro Taima, Yasushi Asami and Kimihiro Hino

Introduction

Recently there has been a policy shift in Japan from the development of new cities to the renovation of existing cities. Particularly, in central commercial zones in Tokyo, the importance of city block restructuring has been strongly emphasized. Three reasons can be identified. First, in a small city block, large buildings cannot be constructed because of urban planning regulations, which may fail to motivate developers to renovate. Second, large offices cannot be located in a small city block. Therefore, the city may lose the opportunity to attract investment from global industries. Third, the existence of roads that are too narrow is an obstacle to disaster preparedness. For all these reasons, city block restructuring has become necessary.

In Japan certain areas have been classified as "urgent urban renewal areas" based on the Urban Renewal Act. Urgent renovation is required in this context. Local governments can relax urban planning regulations such as the floor area ratio (FAR) and motivate developers to renovate. The Ministry of Land, Infrastructure, Transport and Tourism (MLIT) has prepared guidelines for city block restructuring. These show effective examples of city block restructuring and prompt local governments to renovate city blocks. Developers are motivated to renovate city blocks, which can create an attractive city in terms of business, tourism and lifestyle. However, the building shape and location after restructuring are not evident. Spatial images of the city cannot be estimated before renovation.

Our previous study (Taima et al. 2016) classified city blocks by the difference in influential factors of building footprint location (building location of the first floor) and examined whether the city blocks in each class showed predictability of building footprint location. "Predictability" means that the building locations of

M. Taima (✉) · Y. Asami · K. Hino
Department of Urban Engineering, University of Tokyo, 1138656 Tokyo, Japan
e-mail: taima@ua.t.u-tokyo.ac.jp

C. Zhou et al. (eds.), *Spatial Data Handling in Big Data Era*,
Advances in Geographic Information Science,
DOI 10.1007/978-981-10-4424-3_14

each city block are estimated accurately. As a result, a city block comprised entirely of office buildings was shown to be one such class.

Office development is one of the major reasons for restructuring city blocks in Tokyo's commercial zones. Policies and legal systems support the initiative. In the future many offices will be developed in Tokyo's commercial zones.

However, the office shape and location after restructuring are not evident. If the building shape and location could be predicted, and reflected in the restructuring plans, the process would be considerably improved. In addition, environmental influences, such as energy consumption and wind direction, can be estimated. Therefore, the predictability of building development in city blocks is crucial for planning city centers.

In this study, a city block with office buildings only is the focus, and the probability of building coverage for each point on every floor level is visualized to produce a spatial image.

City Planning Regulations in Japan

In this section the major city planning regulations are explained (Ministry of Land, Infrastructure and Transport 2003).

Land Use Zones

In Japan land use zones can generally be categorized into residential, commercial and industrial uses. Twelve categories of land use zones are defined and provide a pattern for land use zoning in each type of urban area. Each land use zone is governed by specifications concerning the use of buildings that can be constructed in the zone.

This study focuses on the category of the commercial zone. Banks, cinemas, restaurants and department stores are constructed in this zone. Residential buildings and small factory buildings are also permitted.

FAR and Building Coverage Ratio

In each land use zone category, maximum floor area ratios (%) and maximum building coverage ratios (%) are defined, and they are shown in Fig. 1. In the commercial zone, the maximum floor area ratio (%) is between 200 and 1300%. The maximum building coverage ratio (%) is 80% in all the areas of the commercial zone (see Table 1).

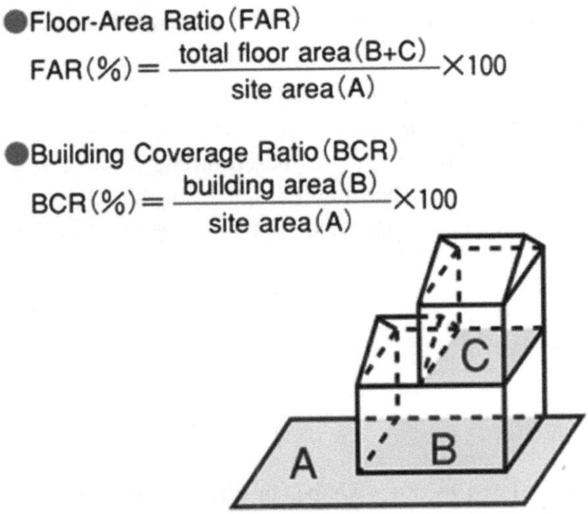

Fig. 1 Floor area ratio and building coverage ratio (Ministry of Land, Infrastructure and Transport 2003)

Table 1 Floor area ratio and building coverage ratio regulations in land use zones (Ministry of Land, Infrastructure and Transport 2003)

Category of land use zone	Maximum floor-area ratios (%)	Maximum building coverage ratios (%}
Category 1 exclusively low-rise residential zone	50 60 80 100 150 200	30 40 50 60
Category II exclusively low-rise residential zone	50 60 80 100 150 200	30 40 50 60
Category 1 mid/high-rise oriented residential zone	100 150 200 300 400 500	30 40 50 60
Category II mid/high-rise oriented residential zone	100 150 200 300 400 500	30 40 50 60
Category 1 residential zone	100 150 200 300 400 500	50 60 80
Category II residential zone	100 150 200 300 400 500	50 60 80
Quasi-residential zone	100 150 200 300 400 500	50 60 80
Neighborhood commercial zone	100 150 200 300 400 500	60 80
Commercial zone	200 300 400 500 600 700 600 900 1000 1100 1200 1300	80
Quasi-industrial zone	100 150 200 300 400 500	50 60 80
Industrial zone	100 150 200 300 400	50 60
Exclusively industrial zone	100 150 200 300 400	30 40 50 60

Restrictions on Building Shape in the Commercial Zone

The restrictions on building shape are different in each zone. In the commercial zone, slant plane restrictions are the major restriction on building shape (Fig. 2). The restrictions limit the building heights based on the distance from the other side of the boundaries of the roads that they face or from the adjacent site boundaries. This ensures adequate space for light and ventilation between buildings or on roads.

Literature Review

Office Shape

Several factors are related to office shape. In office design terms, offices require a high rentable ratio (the rentable area divided by the area available for use in a building) and an efficient working environment. According to Kooijman (2000), office design has changed because of globalization and the needs of the local environment. The office layout is said to be associated with work practices.

Various studies have focused on the analysis of office shape. Chau et al. (2007) analyzed the optimal office shape under building height regulatory restrictions. Shpuza and Peponis (2008) measured the floorplate shape in two different ways and analyzed its influence on office layout. Some studies have proposed models to

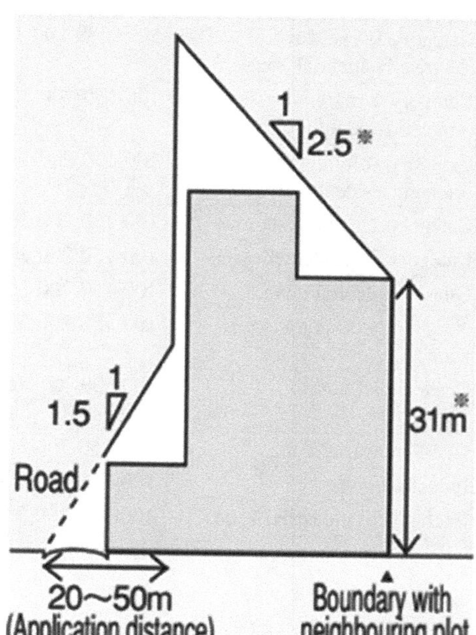

Fig. 2 Slant plane restrictions in commercial zone (Ministry of Land, Infrastructure and Transport 2003)

simulate the optimal office building shape using a genetic algorithm (Grierson and Khajehpour 2002; Ouarghi 2006; Wang et al. 2006).

These studies focused on the office building shape but did not consider the city block shape. The office building shape is influenced by the city block shape, because the city block shape determines almost all the city planning regulations in the block. In this study the office building shape is analyzed from the perspective of the city block shape.

Urban Renewal

Changes in building shape through urban renewal have been analyzed. Some studies have concluded that the initial city block shape affects the building form or city block form after urban renewal. Siksna (1998) examined the influence of the initial city block shape on the building form after development. Ryan (2006, 2008) suggested that new residential patterns are established after the transformation of city blocks.

Processes of urban renewal have also been studied. Lin and Lin (2014) adopted game theory to analyze urban renewal processes based on the characteristics of landowners. In addition, some studies have focused on the process of change in city blocks in the center of Tokyo (Matsukura and Miyawaki 2006), the development process of traditional rectangular city blocks in Kyoto (Hayami 2009) and the development process of city blocks used for office buildings in Marunouchi (Nomura 2014).

These researchers studied the changing building shape and process of urban renewal, but the building shapes of the future could not be estimated. To estimate these, we must develop a model.

Building Location

Malcata-Rebelo and Pinho (2010) found that land use and office location were the relevant variables to the mechanisms of office supply, office demand and market equilibrium. They concluded that these results support municipal decisions concerning office location and management. However, it is not obvious where buildings are located in the new shape of a city block after city block restructuring.

There are many studies about the land use of city blocks. Makio et al. (2006) analyzed apartment house location on the city block scale after a change of building use to apartments. Targeting a provincial city, Saito and Kato (2013) researched changing land use and the current status of each city block. Nam et al. (2007, 2008) analyzed parking based on the relation between green space conservation and business balance. Nagatomi et al. (2007) examined land use in city blocks adjacent to a highway. Nakao and Ito (2012) analyzed urban conditions in terms of building density and building coverage ratio in a city block. Kawaguchi et al. (2015) conducted a quantitative study about the relation between scale and fluctuation of open space and scale and fluctuation of green space. Matsumiya et al. (2014) calculated the distribution of the open space ratio among buildings in a city. These studies analyzed the use of space in city blocks, but further analysis is required to estimate future uses.

Estimation

In this study building location and floor area are estimated. Some estimation methods for urban physical status have been developed in previous research.

Asami amd Ohtaki (2000) developed a model to estimate detached house location. Orford (2010) developed a methodology for estimating the floor area of individual properties from digital infrastructure data, which were, however, deficient in detail. Shiravi et al. (2015) assessed the utility of some models for estimating floor area using three data sources: a geographic vector building footprint layer, a LiDAR data set and field survey data for the south side of the city of Fredericton, Canada. They discussed the reliability and accuracy of each model. In other research Brunner et al. (2009) extended a methodology for building height estimation and tried to improve its accuracy. Schmidt et al. (2010) presented an approach to the estimation of building density on the city block scale.

Many researchers have focused on the estimation of land use: for example, building block use (Spyratos et al. 2016), urban land change (Güneralp et al. 2012) and future urbanization (Debnath and Amin 2016).

Energy and Urban Physical Condition

Estimations of building shape can be applied in many fields. Energy is one such field. Some researchers have analyzed the relation between energy and urban physical condition. Ourghi et al. (2007) developed a method for predicting the impact of the shape of an office building on its annual cooling and total energy. The analysis indicated the strong correlation between the shape of a commercial building and its energy consumption. The result also showed a direct correlation between relative compactness and total building energy use as well as the cooling energy requirement. Rode et al. (2014) examined the theoretical heat energy demand of different types of urban form. They concluded that compact and tall building types had the greatest heat energy efficiency on the neighborhood scale and detached housing had the lowest. Mortimer et al. (2000) studied various aspects of the patterns of energy use in non-domestic buildings derived from the statistical analysis of data.

On the urban scale, Ko and Radke (2014) provided an empirical evaluation of the association between urban form and residential energy use, particularly residential electricity use, for space cooling. The study revealed that urban forms have a statistically significant impact in terms of saving energy for cooling. O'Brien et al. (2010) examined the relation between net energy use and three housing forms: low-density detached homes, medium-density townhouses and high-density high-rise apartments in Toronto. The results show that high-density development uses one-third less energy than low-density development. Only when the personal vehicle fleet or solar collectors are made to be extremely efficient does the trend reverse; low-density development results in lower net energy. These results showed a paradoxical relationship between the density of solar housing and net household energy use.

The other benefit of building shape estimation is that planners can understand the building shapes and locations of the future and use the result for planning. In addition, citizens can easily understand the future image in the city block. They can judge whether the urban renewal plan is better.

Study Area and Data Source

In this study an urban planning GIS data set of Tokyo (March 2013) is used. This data set contains building types, the number of floors, land use zones, the FAR and the building coverage ratio.

City blocks in urgent urban renewal areas are chosen for analysis, because the areas are designated for urgent city block restructuring. The urgent urban renewal area is shown in Fig. 3. The Government finances and promotes the development.

Regarding building location analysis, the difference in the FAR may influence the building location. Therefore, the FAR of blocks is set to be equal to 600%, as this is found to be the mode value of all the blocks in the urgent urban renewal area. As a result, 205 city blocks are chosen, which are used as reference blocks.

Method

Model

A building location estimation (BLE) model was developed to estimate building locations from a city block shape. A similar model has been used to estimate a detached house location (Asami and Ohtaki 2000).

If two city blocks are similar in shape and other spatial features, we can expect that the building locations in the city blocks will tend to be similar. This naive but simple assumption enables us to estimate the building location. Accordingly, a

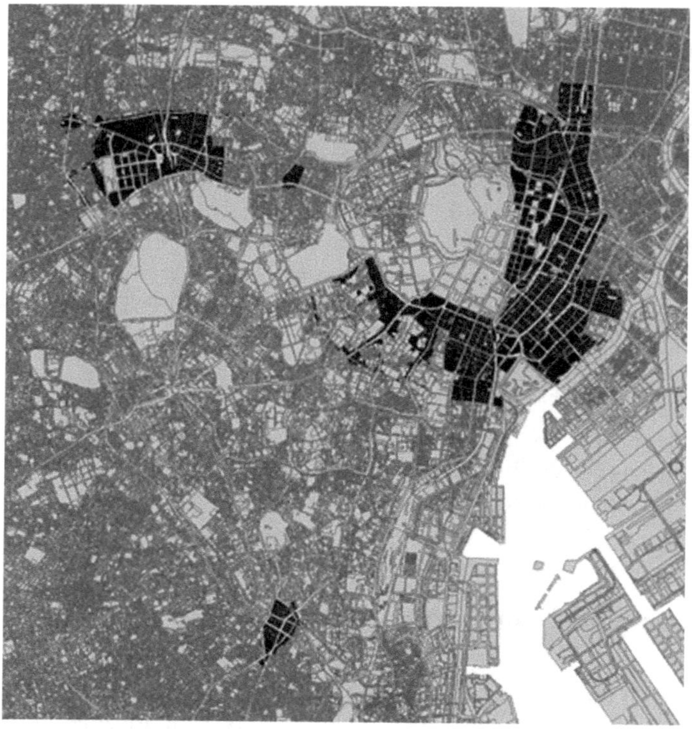

Fig. 3 Urgent urban renewal area in Tokyo (*black area*)

model was developed to estimate the probability of each point on every floor level covered by a building.

More specifically, the locations of buildings on reference blocks were overlaid so that the gravity center of the reference blocks matched that of a given block. A probability was assigned to each overlaid layer depending on the similarity of the block shape. The Lee-Sallee measure (Lee and Sallee 1970) judges the similarity between two city blocks by the quotient value, the intersection area divided by the union area of the two blocks.

An index expressing the similarity between two city blocks, the similarity index, s, is defined below.

Generally, a city block can be treated as a compact set on a two-dimensional plane. Let x. a point on the plane. Let $x(\in X)$ be a point in the city block X and let $g(X)$ be a vector of the gravity center of the city block X. $A(X)$ is the area of the city block X. The similarity index, s, is calculated as the value, that is, the intersection area divided by the union area of the two blocks, by matching the gravity center of the two blocks. Let $G(X)$ be the set that is given by moving in parallel the city block X, so that the gravity center of the block coincides with the origin. Set $G(X)$ is defined as follows:

$$G(X) = \{z : z = x - g(X), x \in X\} \tag{1}$$

Based on the Lee–Sallee measure, the temporal similarity index, s^*, between city block X and city block Y is defined as follows:

$$s^*(X, Y) = \frac{A(G(X) \cap G(Y))}{A(G(X) \cup G(Y))} \tag{2}$$

Building location greatly depends on the direction of any adjacent road. The direction of the road is different in each city block. The influence of the adjacent road on the building location in city blocks is not particularly different when a city block revolves around a gravity center within angle $\pm\pi/4$. Allowing for such revolving, the final similarity index, s, is defined as follows. Let $R(X, \theta)$ be the revolving city block θ around the gravity center of the city block X. The similarity index, s, is defined as follows:

$$s(X, Y) = \max_{-\pi/4 \leq \theta \leq \pi/4} \frac{A(G(X) \cap R(G(Y), \theta))}{A(G(X) \cup R(G(Y), \theta))} \tag{3}$$

Reference blocks are used to calculate the probability of building location. Let I be a set of reference blocks. If two city blocks are similar in shape and other spatial features, we can expect that the building locations in the city blocks will tend to be similar. Therefore, if the building location in city block X is estimated, it is appropriate to use the city block that has a high similarity index, s. It is possible to use the similarity index, s, as a weight that determines the priority of the reference city block. However, if this is the case, then all the reference blocks must be used for the BLE model. A better model can be developed by rejecting city blocks with an apparently different shape. To exclude differently shaped city blocks, it is necessary to designate the weight as zero. To do so, we define the weight function, $f(s)$, based on the similarity index, s, as follows:

$$f(s) = \begin{cases} \frac{s-t}{1-t} & s > t \\ 0 & s \leq t \end{cases} \tag{4}$$

where t is a parameter and its range is zero to one ($t \in [0, 1]$). When the similarity index, s, is less than parameter t, the weight function, $f(s)$, is zero (Fig. 4). The similarity index, s, and weight function, $f(s)$, between city block X and city block Y are expressed as $s(X, Y)$ and $f(s(X, Y))$.

The way to estimate building location can be described as follows. Let $G(i)$ be the reference city block by moving city block $i(\in I)$ in parallel so that its gravity center coincides with the origin, and let B_i be a building location set on the reference city block $G(i)$. In the case in which a point z. $G(i)$ is $z \in B_i$, the building covers point z..hus, the more building location set B_i covers point z.,he more likely point $x(\in X)$ is to be covered. The building existing probability $p(x, X, I)$ (the probability that point x. covered by buildings) is defined by the weight function,

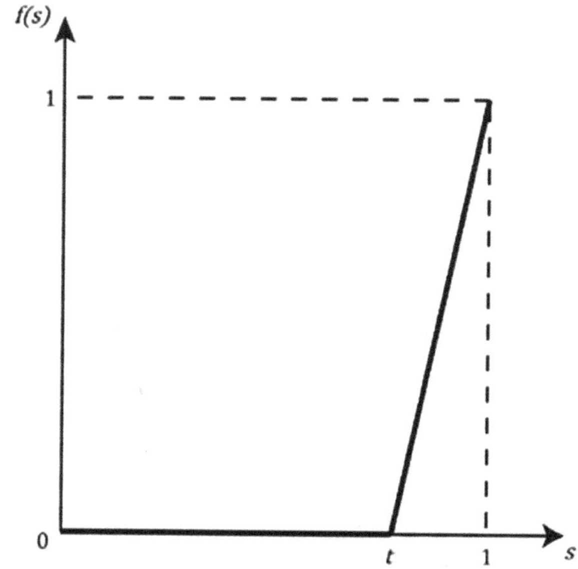

Fig. 4 Weight function, $f(s)$

$f(s)$. First, the indicator function expressing that point x. covered by building location set B_i is defined as follows:

$$\chi(x, B_i) = \begin{cases} 1 & x \in B_i \\ 0 & x \notin B_i \end{cases} \quad (5)$$

Then, the building existing probability $p(x, X, I)$ at point $x(\in X)$ in city block X is defined as follows:

$$p(x, X, I) = \frac{\sum_{i \in I} f(s(X, i))\chi(z, B_i)}{\sum_{i \in I} f(s(X, i))}, \mathrm{x}(\in X) \quad (6)$$

Parameter t is decided so that the accuracy of the building location estimation is maximal. To this end, an arbitrary city block, i, is chosen from the reference city block set, I, and the building location on the city block, i, by using reference city blocks except for city block i (let I_{-i} be the reference city block set except for city block i). We calculate the highest ρ, which is the estimation accuracy index, by trying all of the reference blocks, $i(\in I)$, where the estimation accuracy index, ρ, is defined as follows:

$$\rho = \sum_{i \in I} \int_{x \in B_i} [p(x, i, I_{-i})\chi(x, B_i) - p(x, i, I_{-i})(1 - \chi(x, B_i))]dx \quad (7)$$

The estimation accuracy index, ρ, expresses the summation of both the integral value of the building existing probability at the points covered by buildings and the integral value of the building existing probability at the points not covered by buildings. The value of parameter t is used for the BLE model in the highest estimation accuracy index, ρ.

Influential Factor of Building Location in a City Block

In the future many office buildings will be built in Tokyo, but their location and shape are not evident. One hypothesis is that the locations of each floor of buildings will be estimated accurately (predictable) after the city blocks comprised entirely of office buildings are chosen from all the city blocks. "Predictability" means that the building location of each city block can be estimated accurately by the BLE model.

To determine whether an office building is predictable, the BLE model error is used. The error of the model is calculated as follows. The building existing probability at point $x(\in i)$ in reference city block $i(\in I)$ is expressed as $p(x, i, I_{-i})$. $\chi(x, B_i)$ is an indicator function equal to one when point x. included in B_i and zero otherwise. The error ratio is calculated as the integral value of the absolute value of the difference between $p(x, i, I_{-i})$ and $\chi(x, B_i)$ divided by area $A(i)$ of the reference city block, i. The error ratio is calculated by all the reference city blocks and summed. The error ratio of the estimation of the reference city block set, I, is calculated as the sum divided by the number of reference city blocks, N_I. The error ratio, E, is defined as follows:

$$E = \frac{\sum_{i \in I} \frac{\int_{x \in i} |p(x,i,I_{-i}) - \chi(x,B_i)| dx}{A(i)}}{N_I} \tag{8}$$

The error ratio, E, is used as an index of the model's accuracy. It is used to judge whether the building location of a city block comprised entirely of office buildings can be estimated accurately. If the error ratio, E, of the classified blocks is smaller than that of the unclassified ones, it means that the extracted city blocks have better predictability of the building locations. In the classified city blocks, the building locations can be estimated accurately by the BLE model and the building locations have predictability. Therefore, the class is a factor that influences building locations (influential factor). On the other hand, if the error ratio, E, of the classified blocks is larger than that of the unclassified ones, the building locations in the city blocks are scattered in the classification. The method described above can ascertain whether the building locations of a city block comprised entirely of office buildings can be estimated accurately.

Visualization

The probability of building coverage for each point on every floor level can be used to predict the potential urban environment before the development. In particular, it is important to know office building shapes, because many offices will be in Tokyo's central zones. Therefore, the estimation of the probability of building coverage for each point on each floor level needs to be visualized to obtain the spatial image.

The building existing probability $p(x, X, I)$ is visualized as follows. First, hypothetical blocks are set, and the building existing probability of the points on the block is visualized. Points are set every 1 m in a north-south direction and an east–west direction. Then, reference blocks are overlaid on a hypothetical block. Second, the building existing probability of each point is calculated by summing the building existing probability of the point overlaid by reference city blocks, which is calculated based on the similarity between the reference city blocks and the hypothetical block. The probability is expressed by brightness. In the black area, the probability is high. Conversely, in the white area, the probability is low.

Hypothetical blocks are set by changing their size, and the probability of building coverage for each point on every floor level is calculated. Hypothetical blocks are rectangular and set based on the size 35 m*35 m block. The reasons for the size are based on the mean area and mean perimeter of the reference city blocks. The mean area of all the city blocks is 1285.54 m^2, and the mean perimeter of all the city blocks is 147.41 m. If the city blocks are rectangles, the average shape is calculated as 35 m*35 m from the average area and the perimeter. In all the rectangular blocks, each edge of hypothetical rectangular blocks varies from 20 m to 60 m every 5 m. Almost all the reference city blocks are included in this range. The size of each city block is shown in Tables 2 and 3. In Table 2 the east–west side of the city block varies from 20 m to 60 m, while the north–south side is fixed as 35 m. On the other hand, in Table 3, the north–south side of the city block varies from 20 m to 60 m, while the east–west side is fixed as 35 m. The building existing probability of each hypothetical block is also visualized.

In addition to the visualization of building existing probability, the "estimated area" (the building area considering the probability) and the estimated building area ratio (the ratio of the estimated area in the city block area) are calculated. The estimated area is calculated as the block area multiplied by the average building existing probability of the block, and the estimated building area ratio is calculated as the estimated area divided by the block area.

The index known as the volume sufficiency ratio can measure how far the buildings occupy the maximum volume of the city block. The volume sufficiency ratio is defined as the whole building floor area divided by the maximum volume of the city block. The maximum volume of the city block is calculated as the city block area multiplied by the FAR (600%).

Table 2 Size of each city block (city block ns35*es20–ns35*ew60)

City block	ns35*ew20	ns35*ew25	n&35*ew30	ns35*ew35	ns35*ew40	ns35*ew45	ns35*ew50	n&35*ew55	ns35*ew60
North–South direction side (m)	35	35	35	35	35	35	35	35	35
East–West direction side (m)	20	25	30	35	40	45	50	55	60

Table 3 Size of each city block (city block ns35*es20–ns35*ew60)

City block	ns20*ew35	ns25*ew35	ns30*ew35	ns35*ew35	ns40*e*v35	ns45*ew35	ns50*ew35	ns55*ew35	ns60*ew35
North–South direction side (m)	20	25	30	35	40	45	50	55	60
East–West direction side (m)	35	35	35	35	35	35	35	35	35

Table 4 Error ratio, *E*

Class	Sample number	Error ratio E	Difference
All city blocks	205	0.330	–
City blocks comprised entirely of office buildings	34	0.319	−0.011

Results

Error Ratio, E

According to the result of Table 4, the error ratio, *E*, of the class with city blocks composed of office buildings is smaller than that of unclassified ones (class: all city blocks). In this case the classification of city blocks comprised entirely of office buildings means that the extracted city blocks have better predictability of the building location. In this classification the building locations can be estimated accurately by the BLE model. Therefore, all the office buildings in a city block can be seen as the influential factor of the building footprint location.

Visualization of Building Existing Probability

The building existing probability is visualized in Figs. 5 and 6. The similarity between the reference city block and the hypothetical block is not over the similarity index, *t*, and the building location cannot be estimated by the BLE model. In this case the symbol "–" is marked.

Estimated Area and Estimated Building Area Ratio

The estimated area and the estimated building area ratio of each city block are calculated in Tables 5 and 6. Figures 7 and 8 show the estimated building area ratio.

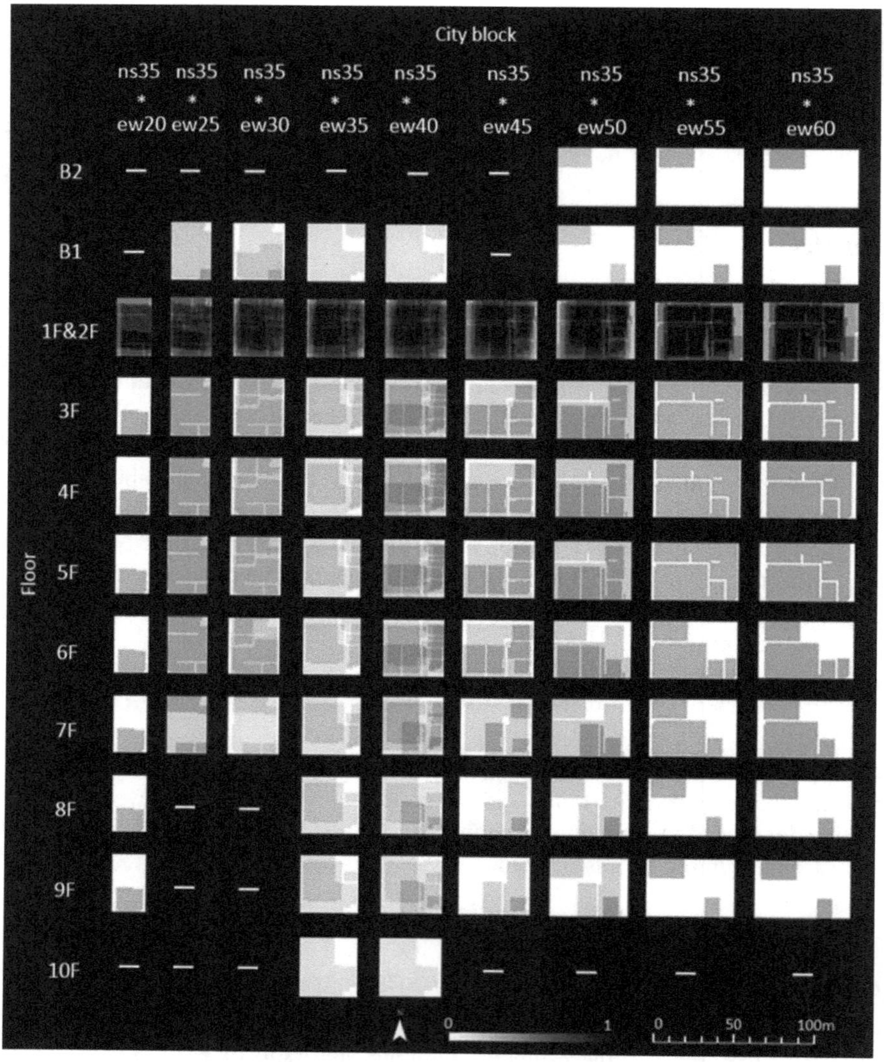

Fig. 5 Visualization of the building existing probability (city block ns35*es20–ns35*ew60)

Volume Sufficiency Ratio

The volume sufficiency ratio is calculated as shown in Tables 7 and 8. The results are also shown in Figs. 9 and 10.

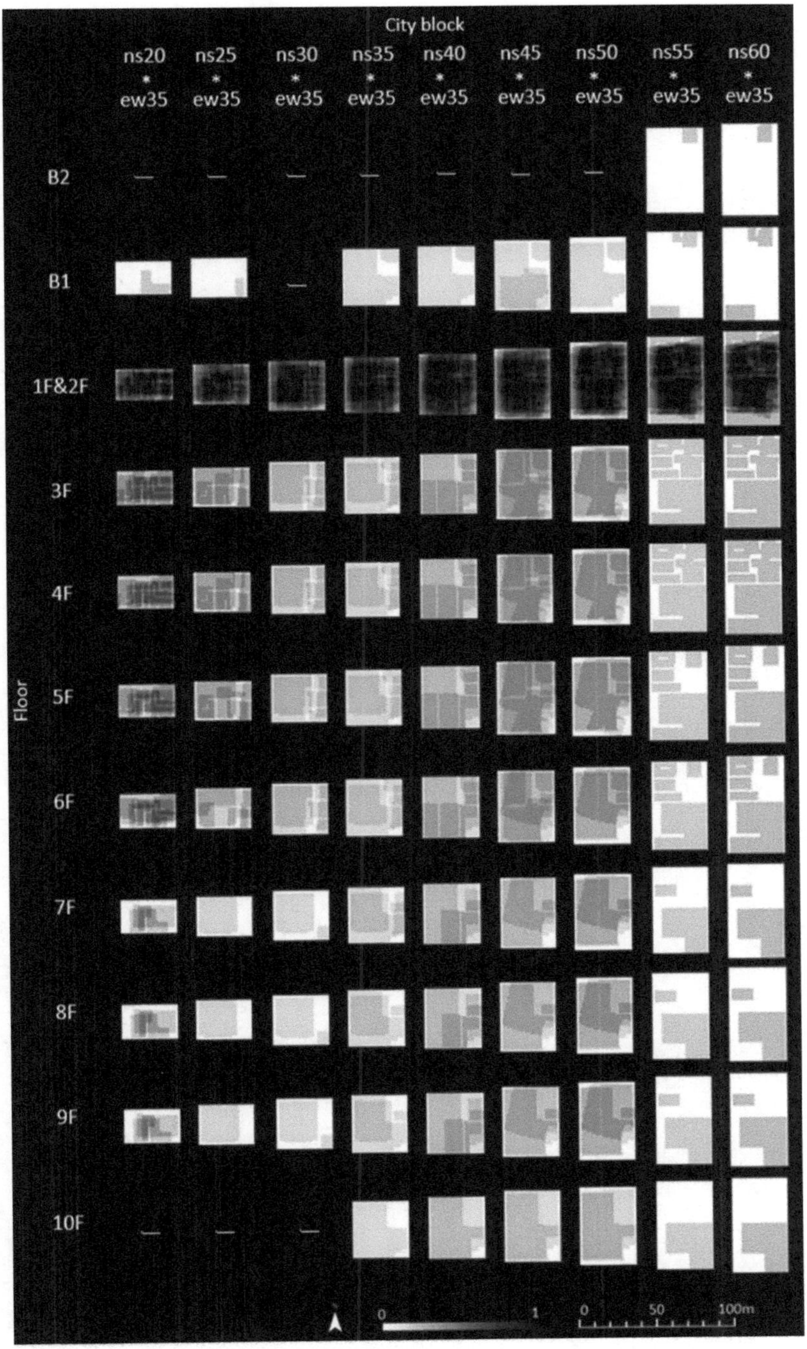

Fig. 6 Visualization of the building existing probability (city block ns20*es35–ns60*ew35)

Table 5 Estimated area and estimated building area ratio (city block ns35*es20–ne6035*ew60)

City block			ns35*ew20	ns35*ew25	ns35*ew30	ns35*ew35	ns35*ew40	ns35*ew45	ns35*ew50	ns35*ew55	ns35*ew60
Area (m^2)			700	875	1050	1225	1400	1575	1750	1925	2100
Floor	B2	Estimated area (m^2)	–	–	–	–	–	–	40.71	62.14	72.42
		Estimated building area ratio	–	–	–	–	–	–	0.02	0.03	0.03
	Bl	Estimated area (m^2)	–	125.2	130.16	107.75	97.86	–	59.03	88.87	103.58
		Estimated building area ratio	–	0.14	0.12	0.09	0.07	–	0.03	0.05	0.05
	1F&2F	Estimated area (m^2)	490.89	637.58	787.31	936.77	984.81	1122.55	1174.5	1377.18	1480.95
		Estimated building area ratio	0.7	0.73	0.75	0.76	0.7	0.71	0.67	0.72	0.71
	3F	Estimated area (m^2)	52.21	232.77	232.53	187.03	342.39	269.89	407.51	410.69	478.67
		Estimated building area ratio	0.07	0.27	022	0.15	0.24	0.17	0.23	0.21	0.23
	4F	Estimated area (m^2)	52.21	232.77	232.53	185.36	340.87	269.89	407.51	410.69	478.67
		Estimated building area ratio	0.07	0.27	0.22	0.15	0.24	0.17	0.23	0.21	0.23
	5F	Estimated area (m^2)	52.21	232.77	232.53	185.36	340.87	269.89	407.51	410.69	478.67
		Estimated building area ratio	0.07	0.27	0.22	0.15	0.24	0.17	0.23	0.21	0.23
	6F	Estimated area (m^2)	52.21	232.77	219.71	185.36	338.25	266.15	328.81	284.93	332.1
		Estimated building area ratio	0.07	0.27	0.21	0.15	0.24	0.17	0.19	0.15	0.16
	7F	Estimated area (m^2)	52.21	170.91	111.65	174.55	294.98	207.21	275.37	263.15	306.71
		Estimated building area ratio	0.07	0.2	0.11	0.14	0.21	0.13	0.16	0.14	0 15
	8F	Estimated area (m^2)	52.21	–	–	157.38	207.49	104.12	159.19	88.87	103.58
		Estimated building area ratio	0.07	–	–	0.13	0.15	0.07	0.09	0.05	0.05
	9F	Estimated area (m^2)	52.21	–	–	157.38	207.49	104.12	159.19	88.87	103.58
		Estimated building area ratio	0.07	–	–	0.13	0.15	0.07	0.09	0.05	0.05
	10F	Estimated area (m^2)	–	–	–	96 94	88.04	–	–		
		Estimated building area ratio	–	–	–	0.08	0.06	–	–	–	–

Table 6 Estimated area and estimated building area ratio (city block ns20*es35 ~ ne60*ew35)

City block			ns20*ew35	ns25*ew35	ns30*ew35	ns35*ew35	ns40*ew35	ns45*ew35	ns50*ew35	ns55*ew35	ns60*ew35
Area (m^2)			700	875	1050	1225	1400	1575	1750	1925	2100
Floor	B2	Estimated area (m^2)	–	–	–	–	–	–		14.91	22.58
		Estimated building area ratio	–	–	–	–	–	–		0.01	0.01
	B1	Estimated area (m^2)	23.82	6.79	–	107.75	118.31	178.61	143.21	45.05	72 07
		Estimated building area ratio	0.03	0.01	–	0.09	0.08	0.11	0.08	0.02	0.03
	1F&2F	Estimated area (m^2)	475.07	567.79	736.87	936.77	1013.97	1058.02	1141.62	1204.37	1279.42
		Estimated building area ratio	0.68	0.65	0.7	0.76	0.72	0.67	0.65	0.63	0.61
	3F	Estimated area (m^2)	298.26	234.51	184.68	187.03	307.48	447.4	434.52	242.47	343
		Estimated building area ratio	0.43	0.27	0.18	0.15	0.22	0.28	0.25	0.13	0.16
	4F	Estimated area (m^2)	287.66	234.51	184.68	185.36	304.89	443.78	430.17	242.47	343
		Estimated building area ratio	0.41	0.27	0.18	0.15	0.22	0.28	0.25	0.13	0.16
	5F	Estimated area (m^2)	269.4	219.86	184.68	185.36	304.89	443.78	430.17	196.43	280.48
		Estimated building area ratio	0.38	0.25	0.18	0.15	0.22	0.28	0.25	0.1	0.13
	6F	Estimated area (m^2)	242.98	174.38	181.61	185.36	304.89	406.68	404.69	196 43	280.48
		Estimated building area ratio	0.35	0.2	0.17	0.15	0.22	0.26	0.23	0.1	0.13
	7F	Estimated area (m^2)	155.5	121.85	168.11	174.55	272.59	310.94	388.92	170.26	237.06
		Estimated building area ratio	0.22	0.14	0.16	0.14	0.19	0.2	0.22	0.09	0.11
	8F	Estimated area (m^2)	102.68	63.65	69.85	157.38	272.59	310.94	388.92	121.73	165.64
		Estimated building area ratio	0.15	0.07	0.07	0.13	0.19	0.2	0.22	0.06	0.08
	9F	Estimated area (m^2)	102.68	63.65	69.85	157.38	272.59	310.94	388.92	121.73	165.64
		Estimated building area ratio	0.15	0.07	0.07	0.13	0.19	0.2	0.22	0.06	0.08
	10F	Estimated area (m^2)	–	–	–	96.94	209.63	227.23	274.78	106.66	145.88
		Estimated building area ratio	–	–	–	0.08	0.15	0.14	0.16	0.06	0.07

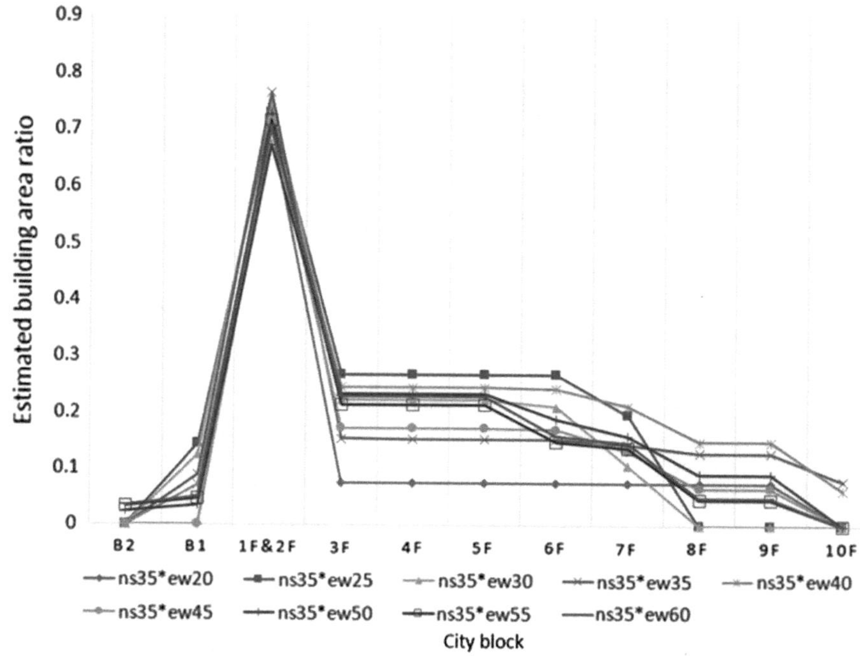

Fig. 7 Estimated building area ratio (city block ns35*es20–ns35*ew60)

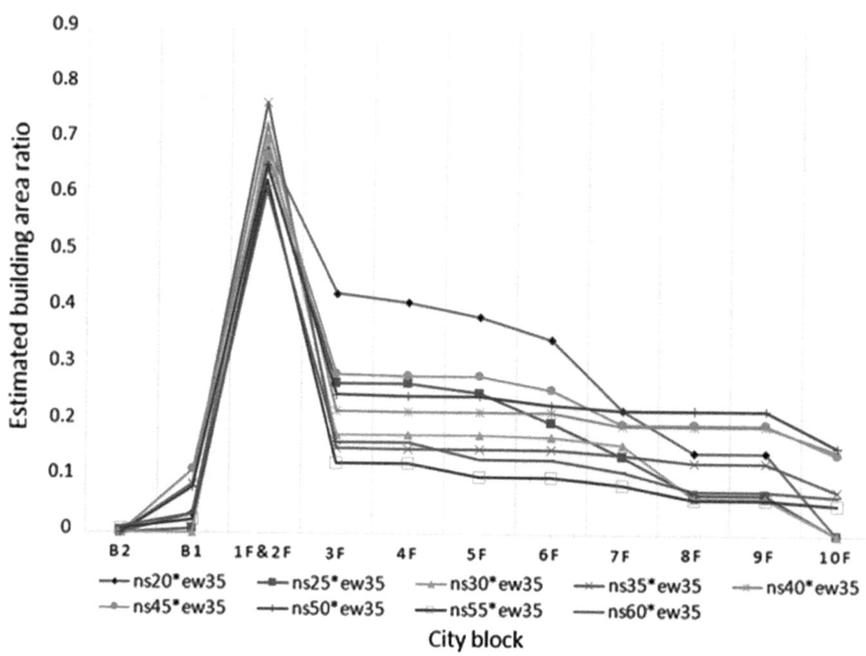

Fig. 8 Estimated building area ratio (city block ns20*es35–ns60*ew35)

Table 7 Volume sufficiency ratio (city block ns35*es20 ~ ns35*ew60)

City block	ns35*ew20	ns35*ew25	ns35*ew30	ns35*ew35	ns35*ew40	ns35*ew45	ns35*ew50	ns35*ew55	ns35*ew60
Volume sufficiency ratio	0.20	0.36	0.31	0.32	0.39	0.28	0.33	0.30	0.31

Table 8 Volume sufficiency ratio (city block ns20*es35–ns60*ew35)

City block	ns20*ew35	ns25*ew35	ns30*ew35	ns35*ew35	ns40*ev*35	ns45*ew35	ns50*ew35	ns55*ew35	ns60*ew35
Volume sufficiency ratio	0.47	0.32	0.28	0.32	0.40	0.44	0.42	0.23	0.26

Discussion

According to Table 4, the error ratio, *E*, of the city blocks comprised entirely of office buildings is smaller than that of unclassified ones (class: all city blocks). Therefore, blocks comprised entirely of office buildings can be seen as an influential factor and provide a steady building location of a city block. Offices require a high rentable ratio (the rentable area divided by the area available for use in a building) and an efficient working environment. These requirements are reflected in the design and may result in predictability of building location in a city block.

According to Figs. 5 and 6, high buildings tend to be developed in large city blocks. In Fig. 5 buildings with more than 8 floors exist in all the city blocks larger than city block ns35*ew35 (including city block ns35*ew35). On the other hand, buildings with more than 8 floors exist in only 1 city block smaller than city block ns35*ew35 (not including city block ns35*ew35). In Fig. 6 buildings with more than 10 floors exist in all the city blocks larger than city block ns35*ew35 (including city block ns35*ew35). On the other hand, buildings with more than 10 floors do not exist in city blocks smaller than city block ns35*ew35 (not including city block ns35*ew35). These results suggest that higher buildings tend to be built in larger city blocks. City planning regulations on the set back and FAR in small blocks cause the buildings to be small. City block ns35*ew35 is the threshold of high buildings.

In addition, buildings with an underground basement tend to be built in large-scale city blocks. In Fig. 5 buildings with a second basement exist in city blocks larger than city block ns35*ew50. In Fig. 6 buildings with a second basement exist in city blocks larger than city block ns55*ew35. The basement is developed to strengthen the foundations of a large city block. Therefore, larger city blocks may promote underground development.

According to Figs. 7 and 8, on the upper floors, the estimated building area ratio is higher in large city blocks. On the other hand, on the lower floors, the estimated building area ratio is higher in small city blocks. In Fig. 7 the estimated building area ratio is high for the third floor (lower floor) in small city blocks like city block ns35*ew25 and city block ns35*ew30, but for the eighth floor (higher floor), city block ns35*ew40, city block ns35*ew35 and city block ns35*ew50 have a high estimated building area ratio. The estimated building area ratio of small city blocks decreases rapidly on higher floors. In Fig. 8 the estimated building area ratio is high for the third floor in small city blocks like city block ns20*ew35 and city block ns25*ew35, but for the eighth floor, larger city blocks like city block ns50*ew35 and city block ns45*ew35 have a high estimated building area ratio. Similar to Fig. 7, the estimated building area ratio of small city blocks decreases rapidly on higher floors. The slant plane restriction is one reason for the results. To promote sufficient space use on higher floors, the city block size should be larger.

According to Fig. 6, the building location may be centered on too small a block. In city block ns35*ew20, the buildings are located at the center of the city block, compared with other city blocks. This is possibly because the regulation of slant

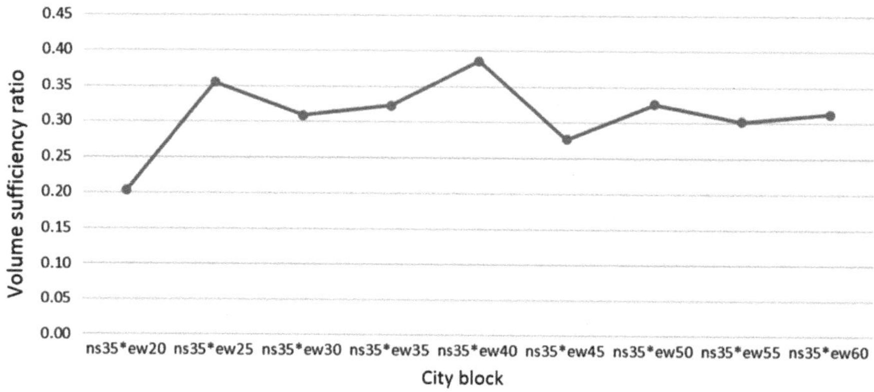

Fig. 9 Volume sufficiency ratio (city block ns35*es20–ns35*ew60)

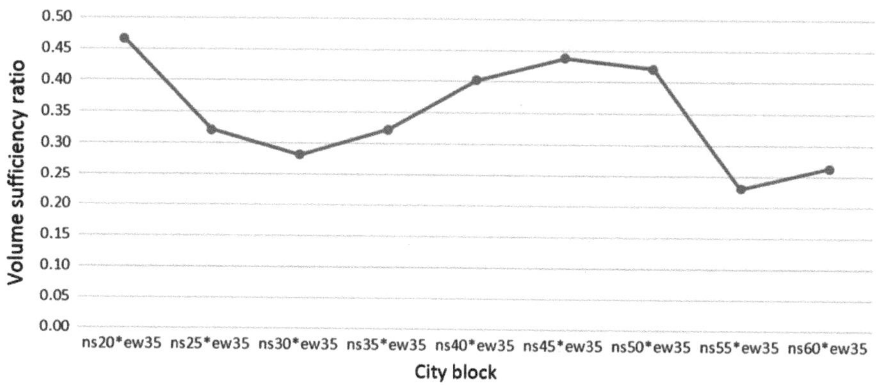

Fig. 10 Volume sufficiency ratio (city block ns20*es35–ns60*ew35)

plane restrictions is strict with regard to small city blocks. Therefore, buildings, particularly high-rise buildings, tend to be located at the center of the city block. In Fig. 5, however, the trends of the building locations cannot be found because of the few reference city blocks of city block ns20*ew35.

Figures 9 and 10 did not show significant results regarding the volume sufficiency ratio. However, this will be discussed below. The volume sufficiency ratio is high around city block ns35*ew40 and city block ns45*ew35, and the volume sufficiency ratio decreases in blocks larger or smaller than city block ns35*ew40 and city block ns45*ew35, except for city block ns20*ew35. It is possible that the buildings do not need to occupy the maximum limited volume in city blocks larger than city block ns35*ew40 and city block ns45*ew35. However, in city block ns20*ew35, the volume sufficiency ratio is high. The buildings occupy much of the volume of the city block. However, city block ns20*ew35 is a small city block. It cannot constitute efficient land use because high buildings cannot be built in a block of this size. Therefore, the city block size should be larger than the size of city block ns35*ew40 or ns45*ew35 to satisfy the needs of the development.

In the future office buildings will be developed and city blocks renovated steadily. The building locations in Figs. 5 and 6 show a visual image of the future of the city blocks. If city blocks of a similar size are developed in the future, the building locations will be similar to the city blocks in Figs. 5 and 6. These results can be used for renovation planning. Previous studies have examined whether energy use is related to building density and floor area and found that energy use can be estimated from the estimation of building locations and the floor area. In addition, the townscape and wind direction can be predicted from the estimation of building locations. In conclusion, the estimation of building locations in city blocks can be applied in various fields.

References

Asami, Y., & Ohtaki, T. (2000). Prediction of the shape of detached houses on residential lots. *Environment and Planning B: Planning and Design, 27,* 283–295.

Brunner, D., Lemoine, G., & Bruzzone, L. (2009). Estimation of building heights from detected dual-aspect VHR SAR imagery using an iterative simulation and matching procedure in combination with functional analysis. *IEEE Radar Conference, 2009,* 1–6.

Chau, K., Wong, S. K., & Yeung, K. C. (2007). Determining optimal building height. *Urban Studies, 44*(3), 591–607.

Debnath, R., & Amin, A. N. (2016). A geographic information system-based logical urban growth model for predicting spatial growth of an urban area. *Planning B: Planning and Design, 43*(3), 580–597.

Grierson, D. E., & Khajehpour, S. (2002). Method for conceptual design applied to office buildings. *Journal of Computing in Civil Engineering, 16*(2), 83–103.

Güneralp, B., Reilly, M. K., & Seto, K. C. (2012). Capturing multiscalar feedbacks in urban land change: A coupled system dynamics spatial logistic approach. *Environment and Planning B: Planning and Design, 39,* 858–879.

Hayami, Y. (2009). The center gutter of Reisen-Chou: Urban dynamics of rectangle block based on the center gutter and neighboring boundary in Kyoto of Edo period. *Journal of Architecture and Planning, 74*(636), 515–522.

Kawaguchi, N., Shimizu, H., Murayama, A., & Takatori, C. (2015). The relation between openness/compactness and scale of green coverage of non-built-up areas and the characteristic feature of their distribution on Nagoya blocks. *Papers on City Planning, 50*(3), 509–516.

Ko, Y., & Radke, J. D. (2014). The effect of urban form and residential cooling energy use in Sacramento, California. *Environment and Planning B: Planning and Design, 41,* 573–593.

Kooijman, D. (2000). The office building: Between globalization and local identity. *Environment and Planning B: Planning and Design, 27,* 827–842.

Lee, D. R., & Sallee, G. T. (1970). A method of measuring shape. *Geographical Review, 60*(4), 555–563.

Lin, Y., & Lin, F. (2014). A strategic analysis of urban renewal in Taipei City using game theory. *Environment and Planning B: Planning and Design, 41,* 472–492.

Makio, H., Sugiyama, S., Tokuono, T., & Nakaniwa, Y. (2006). The increases of apartments and the changes in land uses of street-facing open spaces in extent residential areas. *Journal of Architecture and Planning, 604,* 1–7.

Malcata-Rebelo, E., & Pinho, P. (2010). Evaluation and monitoring of office markets. *Environment and Planning B: Planning and Design, 37,* 305–325.

Matsukura, F., & Miyawaki, M. (2006). Period when streets and blocks were formed in the center of Edo and Tokyo: Historic characteristics of blocks in Tokyo Chuo City and Tsukishima District. *Papers on City Planning, 41*(3), 953–958.

Matsumiya, K., Washizaki, M., Oikawa, K., & Gota, M. (2014). Quantitative analysis of gaps between buildings. *Architecture and Planning, 79*(697), 693–699.

Ministry of Land, Infrastructure and Transport. (2003). *Introduction of urban land use planning system in Japan.* [PDF] http://www.mlit.go.jp/common/000234477.pdf. Accessed June 5, 2016.

Mortimer, N. D., Elsayed, M. A., & Grant, J. F. (2000). Patterns of energy use in nondomestic buildings. *Environment and Planning B: Planning and Design, 27,* 709–720.

Nagatomi, T., Sato, S., & Kobayashi, Y. (2007). Elucidation of the land use condition on the main roadside by building distribution conditions of block units: Main roadside of Oita City, Oita Prefecture. *Papers on City Planning, 42,* 517–522.

Nakao, N., & Ito, K. (2012). Study on aspects of city urban space based on building density on blocks. *Architecture and Planning, 77*(677), 1689–1697.

Nam, T., Sugiyama, S., & Tokuno, T. (2008). The characteristic of apartment houses located in built-up areas from the parking location and building dispositions. *Architecture and Planning, 73*(623), 23–30.

Nam, T., Usami, M., Sugiyama, S., & Tokuno, T. (2007). The parking space installation methods in the apartment housings located in built-up areas. *Journal of Architecture and Planning, 614,* 17–24.

Nomura, M. (2014). Deployment technique in the Marunoucni District analyzed by process of the block formation: Marunouchi District where the Mitsubishi financial clique initiated development vol. 4. *Journal of Architecture and Planning, 79*(698), 1035–1044.

O'Brien, W. T., Kennedy, C. A., Athienitis, A. K., & Kesik, T. J. (2010). The relationship between net energy use and the urban density of solar buildings. *Environment and Planning B: Planning and Design, 37,* 1002–1021.

Orford, S. (2010). Towards a data-rich infrastructure for housing-market research: Deriving floor-area estimates for individual properties from secondary data sources. *Environment and Planning B: Planning and Design, 37,* 248–264.

Ouarghi, R. (2006). Building shape optimization using neural network and genetic algorithm approach. *ASHRAE Transactions, 112*(1), 484–491.

Ourghi, R., Al-Anzi, A., & Krarti, M. (2007). A simplified analysis method to predict the impact of shape on annual energy use for office buildings. *Energy Conversion and Management, 48,* 300–305.

Rode, P., Keim, C., Robazza, G., Viejo, P., & Schofield, J. (2014). Cities and energy: Urban morphology and residential heat-energy demand. *Environment and Planning B: Planning and Design, 41,* 138–162.

Ryan, B. (2006). Morphological change through residential redevelopment: Detroit, 1951–2000. *Urban Morphology, 10*(1), 5–22.

Ryan, B. (2008). The restructuring of Detroit: City block form change in a shrinking city, 1900–2000. *Urban Design International, 13*(3), 156–168.

Saito, A., & Kato, M. (2013). An approach to the change and the actual condition of land use in the block unit: Case study on Taira central city area in Iwaki city. *Journal of the City Planning Institute of Japan, 48,* 315–320.

Schmidt, M., Esch, T., Klein, D., Thiel, M., & Dech, S. (2010). Estimation of building density using terrasar-x-data. In *Geoscience and Remote Sensing Symposium (IGARSS), 2010 IEEE International* (pp. 1936–1939).

Shiravi, S., Zhong, M., Beykaei, S. A., Hunt, J. D., & Abraham, J. E. (2015). An assessment of the utility of LiDAR data in extracting base-year floorspace and a comparison with the census-based approach. *Environment and Planning B: Planning and Design, 42,* 708–729.

Shpuza, E., & Peponis, J. (2008). The effect of floorplate shape upon office layout integration. *Environment and Planning B: Planning and Design, 35,* 318–336.

Siksna, S. (1998). City centre blocks and their evolution: A comparative study of eight American and Australian CBDs. *Journal of Urban Design, 3*(3), 253–283.

Spyratos, S. Stathakis, D., Lutz, M., & Tsinaraki, C. (2016). Using foursquare place data for estimating building block use. *Environment and Planning B: Planning and Design,* forthcoming.

Taima, M., Asami, Y., Hino, K., & Morioka, W. (2016). Influential factors and prediction of location of building coverage in a city block in Tokyo commercial zones. *Theory and Applications of GIS, 24*(2), 85–96.

Wang, W., Rivard, H., & Zmeureanu, R. (2006). Floor shape optimization for green building design. *Advanced Engineering Informatics, 20,* 363–378.

Modelling Urban Growth Evolution Using SLEUTH Model: A Case Study in Wuhan City, China

Wenyou Fan, Yueju Shen, Jianfang Li and Lina Li

Introduction

Urban city is the product of the development of civilization. As a large settlement, it includes the non-agricultural industries and non-agricultural population. Its main elements are population, resources, environment, and it is a complex with great vitality. The history of its development also determines that the urban city will continuously spread around (Dietzel and Clarke 2006). In addition, as a relative open system architecture, ecology and the environment will have some limitations and impact on the urban city in the development process. However, the expansion and changes in the urban city will change the environment nearby, and destroy the ecological structure of the urban city. At different stages of urban development, the overall size of the urban city will continue to change, which may occur at expansion or contraction period. Changes in urban space also show the changes in the functional structure of the city (Clarke et al. 1997; Almeida et al. 2008; Batty and Xie 1994).

W. Fan · Y. Shen (✉) · J. Li
China University of Geosciences, Wuhan 430074, People's Republic of China
e-mail: orangekurt@foxmail.com

W. Fan
e-mail: mapsuv@126.com

J. Li
e-mail: 1184385220@qq.com

L. Li
Tin Yiu Technology Co. Ltd, Beijing Grand, People's Republic of China
e-mail: lilina@greatmap.com.cn

C. Zhou et al. (eds.), *Spatial Data Handling in Big Data Era*,
Advances in Geographic Information Science,
DOI 10.1007/978-981-10-4424-3_15

These are extremely urgent topics for both researchers and government: how to explore the changes in urban space and present the changes visually, how to understand the external and internal regularity of urban expansion better, and how to predict the expansion process and trend of the city in a certain period of time. Therefore, accurately and effectively describing the dynamic development and evolution of urban space, simulating and forecasting different types of urban evolution under different scenarios are conducive to rational and effective planning and adjustment of urban spatial structure and functional composition.

As an analysis and simulation tool of temporal evolution, cellular automata model is based on an individual microscopic derived model system which focused on describing the relationship among the microeconomic factors. It comprises discrete space and time from microscopic to macroscopic. And these characteristics also are proper for describing the urban city's spatial extension. CA Model is a "bottom-up" modeling mechanism model using local changes to show the overall change in a certain extent, and it is quite consistent with the dynamic growth of space research and also is an effective complement to traditional models (Wu and Silva 2010; Xie 2006). In addition, CA model is designed to fit with GIS, and its implementation may well reflect GIS's spatial expression and analytical abilities, such as more visually oriented, more perfect results showed. Further binding with the analysis of GIS, it can more effectively quantify the qualitative results. Consequently, combining GIS with CA model has become an important and scientific support for simulation of the urban growth.

SLEUTH model is a tightly coupled and improved cellular automaton model, which mainly applies to simulating and forecasting the evolution of urban and land use, and it is the core module is the UGM (Clarke Urban Growth Model). This model is designed and developed by Professor Keith Clarke (University of California, Santa Barbara Acronym, UCSB) and his team, which can be used to simulate and analyze the urban growth that is in changeable scale and in global scale (NCGIA 2003). This model is composed of a series of nested cycle of growth rules set in advance and calculates based on the geospatial grid to calculate. It adopts computer simulation to ensure the independent calculation results with the actual results of goodness-of-fit, and then gets the best fitting degree of evolution coefficients group to obtain the closest match so that it could simulate the inertia of the historical evolution coefficient with the subsequent simulation run. The crucial characteristic in SLEUTH model is that it uses self-modifying rules to deduce the state of the urban city of the research areas at different historical stages, and then it does some relative simulations and predictions. Another characteristic of the model that is worth mentioning is that the model regards grid space as an analog unit. It is presented with the pattern of four grid cellular models. As each unit grid is independent, it not only can represent urbanization or non-city values respectively, but also can be limited by the unique cell transformation rules of the model. Finally, the data of the space-time series can be dynamically simulated and analyzed.

Methods

Urban Growth with SLEUTH

Wuhan, as the capital city of Hubei Province, is in the central part of China (Fig. 1). The area of Wuhan is over 8000 km^2, and the population is more than eight million. The city is one of the important cities in China with a long history. This paper selects the major urban areas of Wuhan as study area. We use the SLEUTH model to simulate urban dynamics of Wuhan city. The specification of the SLEUTH model, which is an urban growth based on CA model, has originally been described by Keith Clarke.

SLEUTH Model

The English short name of SLEUTH model comes from the first letter of the name of each input layer: Slope, Land use, Exclusion, Urban Extent, Transportation and Hill shade. It does not require Land use layer only if to simulate urban expansion

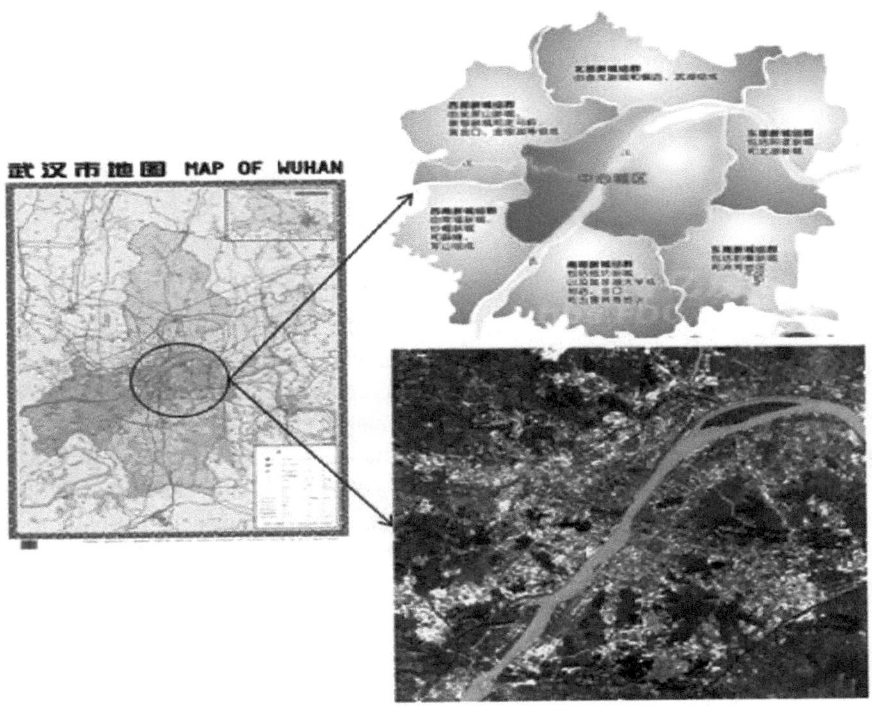

Fig. 1 Location of study area

(Yang and Lo 2003). The required input data of the model must be image format of gray Graphics Interchange Format. Format uniformity is important for geographic data, and it is required for the model to have a consistent projection coordinates, map range, resolution and so on. Moreover, the input name of model data also requires a certain format specification. In the model of SLEUTH, the dynamic urban growth is expressed according to four growth rules: spontaneous growth, new spreading center (diffusive) growth, edge (organic) growth and road influenced growth. Spontaneous growth simulates the occurrence of urban settlement in a new area without pre-existing urban areas and infrastructures, and new spreading center controls the likelihood that a spontaneous growth will affect a center of continued urban growth (Ding and Zhang 2007). Edge growth includes urban growth that occurs outward from city as well as urban infilling. Road influenced growth simulates the tendency of new settlements to appear along transportation lines and encourages urbanized cells to develop along the transportation network. These growth rules are controlled by five growth parameters, such as diffusion, reproduction, extension, road gravity and slope resistance. Each of the parameter has a range of 0–100, which is dimensionless and shows the importance of the corresponding parameter. The diffusion factor determines the overall outward dispersive nature of the distribution; a breed parameter specifies how likely a newly generated detached settlement is to begin its own growth cycle; how much a spread parameter controls diffusion expansion occurring from existing settlements; a slope resistance factor demonstrates the likelihood of settlement extending up steeper slopes and the road gravity factor attracts new settlements toward and along the roads, for example (Silva and Clarke 2002; Clarke 2008).

Procedures

Study Area

Wuhan, known as one of the most important cities in central China, is the capital city of Hubei Province and is also a testing ground for the development of national strategies (Fig. 1). As the largest city of central region, it is known as the "Chicago of the East". Since the Yangtze River runs through the city, Wuhan also known as the "River City". Wuhan city, in the form of radiation, with its many waterways and convenient transportation, enjoys advantageous position making it an important hub in China. The area of Wuhan is over 8000 km^2, and the resident population in Wuhan is more than eight million. This study selects the major urban areas of Wuhan as the study region.

Locating at 29° 58′–31° 22′ N and 113° 41′–115° 05′ E, Wuhan is in the central region of China. It covers about one hundred and thirty-four kilometers from the east to the west, and covers about one hundred and fifty-fifteen kilometers from the north to the south.

In the aspects of economy, at the end of 2014, the total economic output of Wuhan has exceeded one trillion. The local GDP had been rising rapidly from 2010 to 2014. The scale was rapidly rising, and the central strategic fulcrum was increasingly becoming significant. Financial strength continued to be strong. With the continuous growth of tax-revenues, non-tax revenue, and the total public budget revenues, current situation is excellent.

In the aspects of population, the population of Wuhan was over ten million in the end of 2014. Household population was more than eight million, of which agricultural population was two hundred and fifty million and non-agricultural population was five hundred and fifty million. The birth rate was higher than the mortality rate. The net migration rate of population was low, natural population growth rate is relatively stable.

Data and Methodology

SLEUTH model requires five input layer: Slope, Exclusion, Urban Extent, Transportation and Hill shade, respectively. Using RS and GIS, ENVI5.0, ArcGIS10.2, this experiment selected suitable remote sensing image data and digital elevation data(DEM) of Wuhan in 1991, 2000, 2007, 2014 traffic vector data (shape format) in 2015 to simulate. Besides, the experiment takes the administrative division into consideration. The Landsat TM remote sensing image data was used to extract the data of Urban Extent layers, Exclusion layers, and Transportation layer. DEM (digital elevation data) is processed by using ArcGIS. Table 1 shows the data resource of the experiment.

According to the operation requirements of the model, the above data had done some relative treatments to obtain the desired layer. With the support of ENVI5.0, the images of 1991, 2000 and 2007 were geometrically calibrated according to the images of 2014. The image was cut with the vector data to obtain the study area image. Firstly, with the support of ENVI5.0, the images of 1991, 2000 and 2007 were geometrically calibrated according to the images of 2014. The image was cut with the vector data to obtain the study area image. Secondly, each year's traffic layer was made respectively on the basis of remote sensing image and road traffic data of Wuhan. Finally, the slope map and hill shade layer of study area can be made with ArcGIS according to DEM data of Wuhan. All of the image data and

Table 1 Data resource for experiment

Data type	Years	Resource name	Accuracy
Raster	1991, 2000, 2007, 2014	Landsat TM	30*30
Vector	2015	Wuhan administrative divisions	——
	2015	Wuhan transportation data	——
DEM	——	GDEMV2 30 M	30*30

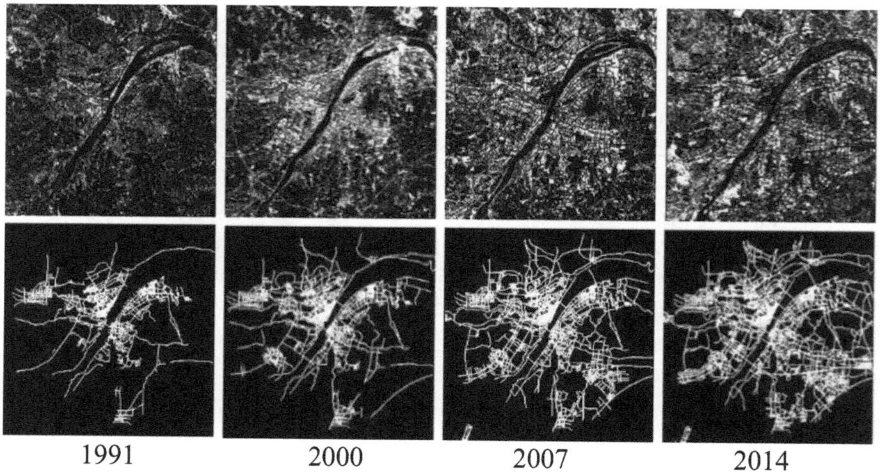

Fig. 2 The extent of Wuhan City and the associated road networks from 1991 to 2014

vector data are unified to set Transverse Mercator (UTM WGS 1984 49 N). Figure 2 shows the extent of Wuhan City and the associated road networks from 1991 to 2014.

Model Calibration

SLEUTH model adopted the Monte Carlo method to do iterative calculation, and the optimal parameters of the model can be determined by the simulation results of different parameters and the fitting degree of the real data (Xi et al. 2009). The model processes and generates thirteen evaluation indicators of the model, which was used to evaluate the experimental results of the model. If the evaluation indicators were used excessively, the behavior of model evaluation will become more complex and the results of the behavior of the model cannot be well reviewed. Collecting and studying the relative literature data in the early period, we can conclude that Lee-Salle shape index is the criteria which is widely adopted in many applications in SLEUTH model. The Lee-Salle shape index of the model is the fitting index in shape between the extended range of the city and the actual range of the city every year. The experiment has also been repeatedly verified. The results using Lee-Salle shape index evaluation are relatively stable. Accordingly, this study chooses Lee- Salle shape index to check the results.

According to the multiple simulation test and literature review, the parameters of the model are set as following the value ranges from 0 to 100, the step increment is set as 25, and Monte-Carlo iterations is set as 4 (MONTE_CARLs O_ITERATIONS = 4) in the coarse calibration of SLEUTH model. The parameter range of Precision calibration and step size are obtained by the previous coarse calibration, and this step will set the different value 8 times to simulate the outcome of the experiment.

Therefore, the experiment selects the final result as the optimal combination of input parameters. Meanwhile, the Monte-Carlo iteration number is set as 100. The optimal parameters for the historical evolution of urban expansion in Wuhan include that the diffusion coefficient is 9, the breed coefficient is 100, the spread coefficient is 75, the slope coefficient is 6, and the road gravity coefficient is 90.

Model Prediction

This paper obtains the optimal parameters which were originated from simulating the historical evolution by extracting the parameters of calibration module. The prediction module model simulates the urban growth trend of Wuhan from 2015 to 2034 which is based on the simulated historical evolution in Wuhan from 1991 to 2014. The SLEUTH provides a simulation environment to explore the consequences of policies taken by decision makers. In this research, we simulated Wuhan City under three scenarios: historical urban growth (scenarios A) which allowed urban expansion without any limitation and a continuation of the historical trend, the environmental protection (scenarios B) and the urban growth limited according to the environmental considerations with slope (scenarios C).

Result and Discussion

Historical Urban Growth

The correction of the model is the simulation result of the historical evolution of urban expansion. To a certain extent, the determination of the coefficient shows the expansion of the situation of Wuhan from 1991 to 2014. The analyses based on behavioral factors are as follows:

1. The diffusion coefficient of model is applied to the spontaneous growth of the rules, and it indicates the probability of the expansion of non-urban area compared to urban areas. That is to say the diffusion coefficient is greater. Besides, it also shows that the city is constantly extended to some extent. In contrast, the diffusion coefficient is smaller, which reflects the development of the city will relatively focus on urban areas to some extent. The diffusion coefficient is 9 which is what we get from model correction. It is relatively small. The urban fringe did not show too much extended trend in the process of city expansion from 1991 to 2014. However, it was relatively stable, and there was no obvious new extension center generated in 1991 and 2014, which may be related to the selected area. The main city of Wuhan is already well-developed, there will not

be no much room for expansion. According to the urban extent layer of study area, Wuhan City has experienced a period of rapid development of urbanization from 1991 to 2014. Since the founding of the People's Republic of China, the level of urbanization of Wuhan has gradually improved. Meanwhile, its development also presented an unbalancing situation. For example, Urban Land-Use increased much faster, and the advantages of its location and convenient transportation made the city bear less resistance in the process of expansion. However, Wuhan, as a city owning so much water, made its urban expansion focusing on three towns into a situation of tripartite confrontation. Consequently, the space is much more fragmentary.

2. Propagation coefficient reached the maximum value of 100, and the spreading coefficient and the road gravity coefficient respectively reached the value 75 and 90. This indicated that the spread rate of urban areas was at a relatively high level from 1991 to 2014. Under the circumstances of less extensible land, urban areas, together with the road, continue to spread to non-proliferation in urban areas together with the road. However, with the influence of road gravity, the government continued to develop construction of the city around the road which had not been urbanized.
3. Slope coefficient is 6, which shows less terrain constraints of Wuhan to a certain extent. The slope maintained low but in constant stability in the range of the study area. The territory is flat in Wuhan. On the other hand, with the driving force of social development, scientific development and technological development, the limitation of terrain has been erased to a certain extent.

Figures 3, 4 and 5 show simulation results between 2015 and 2034, and the growth of urban fringe is increased simultaneously with the urban area. However, it has slowly paced down after 2020, and the growth rate has slowed down. Consequently, we can conclude that the degree of urbanization will become sustainable in the next 20 years, and the urbanization will tend to become saturated in the near future. According to this figure, we can see the growth rate of urbanization will constantly declines in 2018, and finally tend to slow down in 2030. With less

Fig. 3 The increased urban area of Wuhan City by 2034 under scenarios A

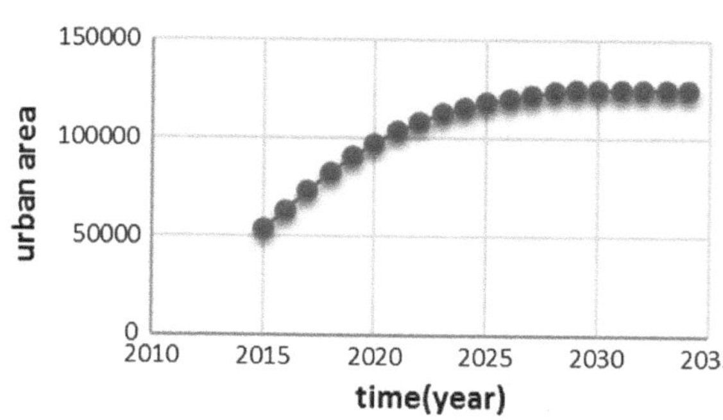

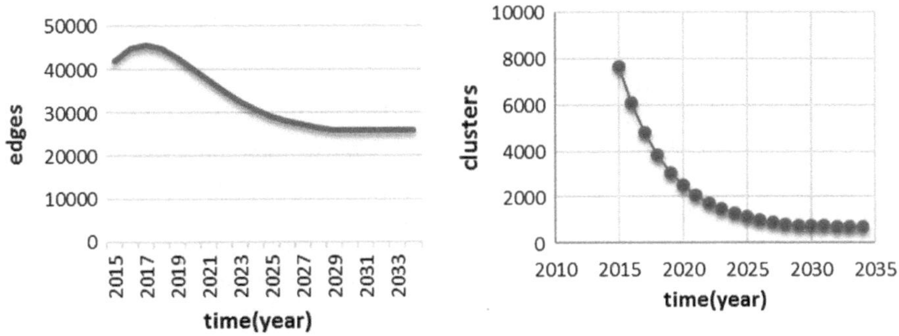

Fig. 4 The increased urban edges and clusters of Wuhan by 2034 under scenarios A

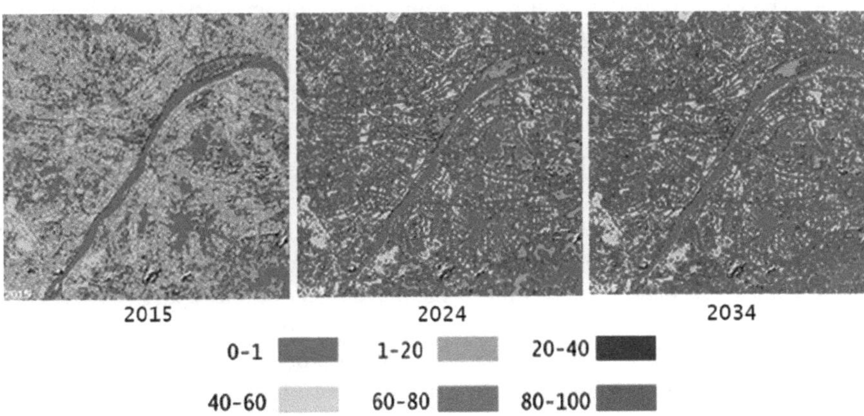

Fig. 5 The urban extent of Wuhan City from 2015 to 2034 under scenarios A

land area that can be exploited, the land away from the road is far more difficult to develop. The slowdown of overall urban construction is inevitable.

In addition, Fig. 4 shows the line charts of edges growth and cluster growth. The effect of margin growth and clustering reflect an ability that a city radiates and absorbs from its surrounding areas. According to line chart of edges growth in urban, the edge growth of urban initially developed rapidly. As time passes by, however, it begins to decline around 2018 and remains flat until around 2028. With the possible reasons of the gradual saturation of urban development, the central urban land of Wuhan around 2018 is already saturated. Furthermore, the rate of development is declining year by year, indicating that the rapid expansion of the city before the study has resulted in the growth of space within the critical state. The result of the transformation of a large number of other types of land into cities in 2018 is that the rate of urban expansion in the decade from 2024 to 2034 is much less than in the previous decade. In addition, according to clustering index line graph in Fig. 4, we can know that the aggregation effect in Wuhan has been a downward trend, and ultimately is going to reach a low clustering value in 2027. It will be maintained a low clustering value ever since. In these images, the higher the value, the more likely the urbanization is. In order to interpret these continuous values more easily, they may be classified by range with different color. What the possible color is a color value in hexadecimal indicates a probability range.

That the three towns of Wuhan are developing independently possibly causes this unique situation led to the development of the lack of a sense of centrality. The three towns in Wuhan have their own center, but none of them can eventually grow into a center which could be common recognized by the whole city. Especially in recent years, competition in the commerce and services is rapidly increasing, and the commercial core in Wuhan is gradually moving from the old center to the outside. The result is that the hollow phenomenon within the inner region is severe, and the city lacks basic service function. In addition, the urban development in Wuhan is inevitably restricted by the natural geographical factors. Being called "the city of hundreds of lakes", with sufficient lake recourse, the GDP of lakes tour in Wuhan is lower than other cities' lakes. Rivers, lakes, low-lying wetlands will cut other complete land into broken blocks, and the city's development space relative to other inland cities is more fragmented. Natural geography forms Wuhan's unique urban development pattern, which provides abundant substantive cases for Chinese urbanization research. On the other hand, it is very regrettable that we have seen these natural geographic features limited the expansion of urban space. The relative dispersion of fragmented land resources is difficult to provide sufficient space for the long-term optimization of cluster type development, which causes that the boundaries of Wuhan has difficultly in extending outwardly.

Based on this, from the perspective of sustainable development, the government must formulate relevant policies to avoid excessive land-occupation, such as protecting the natural ecological environment in the region, earnestly implementing the scientific concept of development. At the same time, we need to improve environmental awareness and raise the coverage of the plants on the earth, protect water bodies and control disordered expansion of the city. Secondly, the balanced development of three towns has always been the basic idea of spatial development of Wuhan and previous overall urban planning. The construction around the Yangtze River Bridge and the ring roads further tighten the relationship of three towns, but there are still large differences in the development foundation, development environment and development stages between three towns. It is inappropriate to bind the three towns together to achieve the true development. The government should attach great importance to the natural barrier dividing of Yangtze River to ensure smooth docking of three town in all levels based on their respective functions. It breaks the status of the three towns that is evenly distributed on the function in spatial layout. Meanwhile, the government should allocate the overall layout of residential, commercial, ecological and other functions.

Scenario Comparison

The scenarios compared to the historical evolution shows there is no doubt that the expansion rate of Wuhan is greatly reduced when intensity of the extension is limited. Compared to the same period of the left image, the scene of Limited-expansion factor will reflect that there are more lands being left, which could leave more space for ecological construction and contributing to protecting the environment. The second scenario keeps the same effect with the scenario of historical evolution, and the expanded form of Wuhan has not changed much. According to this, it is not difficult to see that the terrain that limits its development is not the determining factors for the part of whole city. Consequently, it is necessary to do some corresponding rational planning that contributes to the development and construction of Wuhan. Figures 6 and 7 show the comparison among three Scenarios.

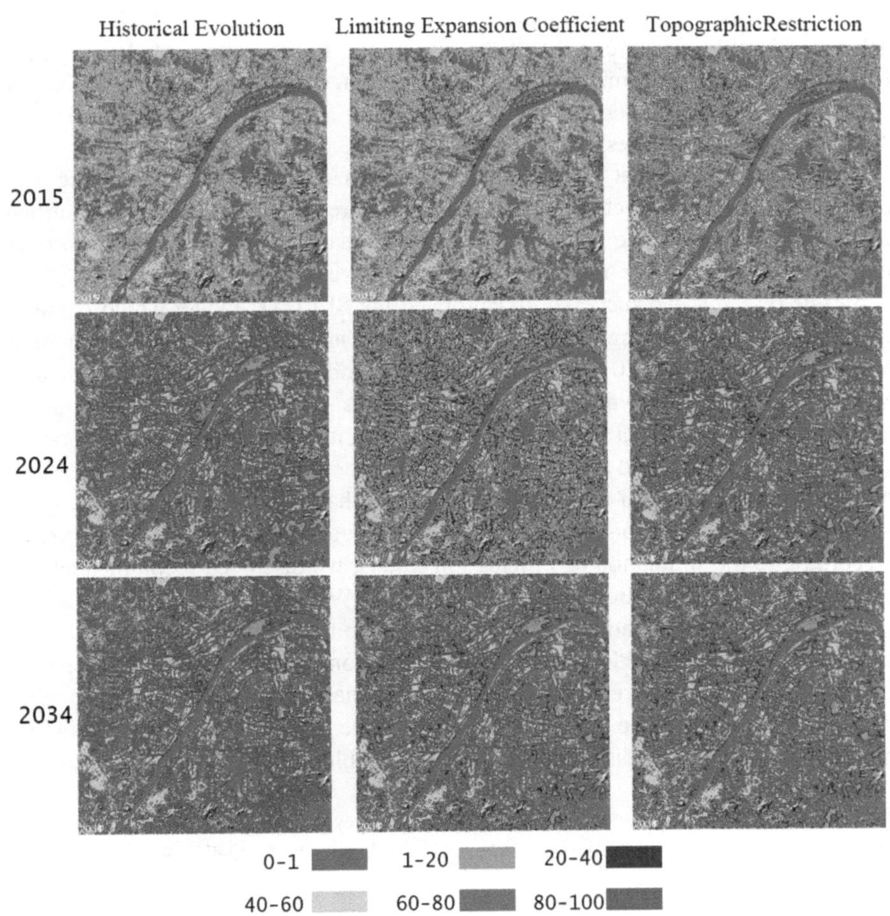

Fig. 6 The urban extent of Wuhan City from 2015 to 2034 under three scenarios

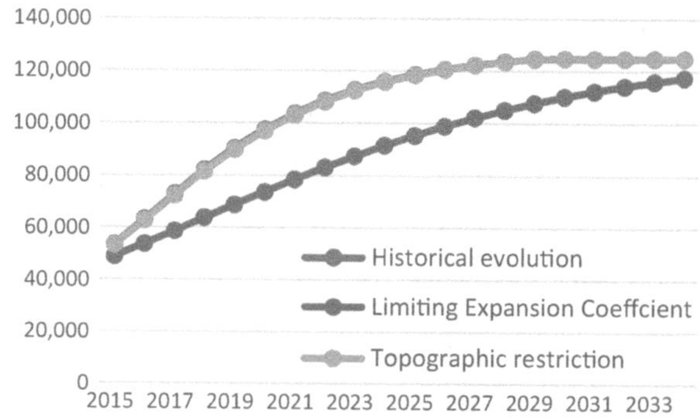

Fig. 7 Predicted urban areas under three Scenarios

Conclusions

SLEUTH model, as a CA based scientific method, combined with GIS and its own advantages, has a unique advantage for the urban growth research. In order to achieve scientific macro control of the city in the future pattern of development trend, we can design some relative scenes to carry out urban planning. In this regard, the model is a good scientific supplement for policy-makers to reach the auxiliary support purposes.

In this paper, we select Wuhan as the study area, and select remote sensing data of many years, road vector data in the urban area, etc. To explore the urban expansion and predict the extended evolution of main urban areas in Wuhan over the next two decades, we raise the corresponding recommendations.

In this research, we successfully calibrated the SLEUTH model which is based on historical data covering the time of 1991–2014 and predict the urban growth in Wuhan within the next 20 years. The SLEUTH model was successfully applied to Wuhan urban space growth study, and experiments show that the SLEUTH model has a good feasibility for application in Wuhan city. Through model calibration procedure, we determined the optimal combination of parameters. According to the city's historical evolution from 1991 to 2014, we have done quantitative analysis, simulation and prediction the urbanization of Wuhan over the next 20 years. According to the simulation and prediction results, we discussed recommendations by taking into consideration the circumstances of Wuhan.

On the basis of simulation studies, by setting up two scenarios respectively, the study area was forecasted again, and the prediction simulation results were analyzed. It is a good reference to bind model scenarios to forecast city planning decisions. The model also shows its ability to serve as a decision support tool and help the managers to realize the outcome of possible actions they might take.

Acknowledgements This research was supported by the Key Project in the National Science and Technology Pillar Program during the Twelfth Five years Plan Period of China (No. 2014AA123001).

References

Almeida, C. M., Gleriani, J. M., Castejon, E. F., & Soares-Filho, B. S. (2008). Using neural networks and cellular automata for modeling intra-urban land-use dynamics. *International Journal of Geographical Information Science, 22*(9), 943–963.

Batty, M., & Xie, Y. (1994). From cells to cities. *Environment and Planning B: Planning and Design, 21*(7), 531–548.

Clarke, K. C. (2008). A decade of cellular urban modeling with SLEUTH: Unresolved issues and problems, Ch. 3 in planning support systems for cities and regions. In R. K. Brail (Ed.), *Lincoln institute of land policy* (pp. 47–60). Cambridge.

Clarke, K. C., Hoppen, S., & Gaydos, L. (1997). A self-modifying cellular automaton model of historical urbanization in the San Francisco Bay area. *Environment and Planning B: Planning and Design, 24*(2), 247–261.

Dietzel, C., & Clarke, K. C. (2006). The effects of disaggregating land use categories in cellular automata during model calibration and forecasting. *Computers, Environment and Urban Systems, 30*(1), 78–101.

Ding, Y., & Zhang, Y. (2007). The simulation of urban growth applying SLEUTH CA model to the Yilan Delta in Taiwan. *Journal Alam Bina, 9*(01), 95–107.

NCGIA. (2003). Project Gigalopolis, NCGIA. http://www.ncgia.ucsb.edu/projects/gig/

Onsted, J., & Clarke, K. (2011). Using cellular automata to forecast enrollment in differential assessment programs. *Environment and Planning B, 38*(5), 829–849.

Onsted, J., & Clarke, K. C. (2012). The inclusion of differentially assessed lands in urban growth model calibration: A comparison of two approaches using SLEUTH. *International Journal of Geographical Information Science, 26*(5), 881–898.

Silva, E. A., & Clarke, K. C. (2002). Calibration of the SLEUTH urban growth model for Lisbon and Porto, Portugal. *Computers, Environment Urban System, 26*(6), 525–552.

Wu, N., & Silva, E. A. (2010). Artificial intelligence solutions for urban land dynamics: A review. *Journal of Planning Literature, 24*(3), 246–265.

Xi, F., Hu, Y., He, H., Wu, X., & Yu, J. (2009). Simulate urban growth based on RS, GIS, and SLEUTH model in Shenyang-Fushun metropolitan area north-eastern China. *Joint Urban Remote Sensing Event*, 1–10, doi:10.1109/urs.2009.5137630

Xie, C. (2006). *Support vector machines for land use change modeling*. UCGE Reports: Calgary.

Yang, Q., Li, X., & Shi, X. (2008). Cellular automata for simulating land use changes based on support vector machines. *Computers & Geosciences, 34*(6), 592–602.

Yang, X., & Lo, C. P. (2003). Modelling urban growth and landscape changes in the Atlanta metropolitan area. *International Journal of Geographical Information Science, 17*(5), 463–488.

FSC
www.fsc.org
MIX
Papier | Fördert
gute Waldnutzung
FSC® C083411

Kolibri 360 GmbH
Ferdinand-Jühlke-Straße 7
99095 Erfurt, Deutschland
produktsicherheit@kolibri360.de